Apple
Watch®

for
dummies®
A Wiley Brand

Apple Watch®

3rd Edition

by Marc Saltzman
Freelance journalist, author, speaker,
and radio and TV personality

A Wiley Brand

Apple Watch® For Dummies®, 3rd Edition

Published by: **John Wiley & Sons, Inc.,** 111 River Street, Hoboken, NJ 07030-5774, www.wiley.com

Copyright © 2020 by John Wiley & Sons, Inc., Hoboken, New Jersey

Published simultaneously in Canada

For general information on our other products and services, please contact our Customer Care Department within the U.S. at 877-762-2974, outside the U.S. at 317-572-3993, or fax 317-572-4002. For technical support, please visit https://hub.wiley.com/community/support/dummies.

Wiley publishes in a variety of print and electronic formats and by print-on-demand. Some material included with standard print versions of this book may not be included in e-books or in print-on-demand. If this book refers to media such as a CD or DVD that is not included in the version you purchased, you may download this material at http://booksupport.wiley.com. For more information about Wiley products, visit www.wiley.com.

Library of Congress Control Number: 2019952387

ISBN: 978-1-119-65866-5 (pbk); ISBN: 978-1-119-65864-1 (ePDF); ISBN: 978-1-119-65868-9 (ePub)

Manufactured in the United States of America

V10015080_102519

Contents at a Glance

Introduction 1

Part 1: Getting to Know Apple Watch 7
CHAPTER 1: Watch This: Introducing Apple Watch 9
CHAPTER 2: Time Out: Setting Up Your Apple Watch 33
CHAPTER 3: Control Freak: Mastering Apple Watch's Interface and Apps 57

Part 2: Just the Tasks, Ma'am! 85
CHAPTER 4: It's About Time: Learning How to Set Watch Faces, Alarms,
         Timers, and More 87
CHAPTER 5: Keep in Touch: Using Apple Watch for Calls, Messages, Texts,
         Walkie-Talkie, Emails, and More........................... 123
CHAPTER 6: In the Know: Staying Informed with Apple Watch................. 161

Part 3: It's All in the Wrist 193
CHAPTER 7: Siri Supersized: Gaining the Most from Your Personal Assistant..... 195
CHAPTER 8: Fitness Fun and Happy Health: Apple Watch Is Your Workout
         Buddy and Digital Doctor 215
CHAPTER 9: Mucho Media: Managing Your Music, Movies, and More 253
CHAPTER 10: Pay for Play: Making Mobile Payments with Apple Watch.......... 277

Part 4: More Apple Watch Tips and Tricks 295
CHAPTER 11: App It Up: Customizing Apple Watch with Awesome
         Apps and More 297
CHAPTER 12: Extra! Extra! Having Fun with Apple Watch 317

Part 5: The Part of Ten 333
CHAPTER 13: Ten Cool Things to Do with Your Apple Watch 335

Index 349

Table of Contents

INTRODUCTION 1

About This Book. 1

How to Use This Book. 2

Foolish Assumptions. 4

Icons Used in This Book 4

Beyond This Book 5

Where to Go from Here 6

PART 1: GETTING TO KNOW APPLE WATCH 7

CHAPTER 1: **Watch This: Introducing Apple Watch** 9

Exploring the Different Apple Watch Collections 10

Figuring Out What Apple Watch Can Do 12

Watch faces 13

Timers and alarms. 13

Caller ID, or even calls. 13

Walkie-Talkie 13

Emergency SOS 13

Text messages 14

Email 14

Wrist-to-wrist communication. 14

Dock. 14

Calendars 14

Maps 16

Siri 16

Fitness. 16

Music connectivity and more. 16

Apple Pay 16

Other apps 18

Other functions 18

Bonus tips. 18

Determining What You Need for Your Apple Watch 18

Getting to Know the Apple Watch Home Screen 20

Learning About Apple Watch's Parts. 24

Watch face 24

Digital Crown button. 25

Side button. 25

Back sensors/charger 26

Watch band 26

Using Apple Watch's Touchscreen 26
Understanding Apple Watch's Wireless Functions
and Internal Sensors...27
    Bluetooth 4.0 27
    Wi-Fi...28
    Cellular 28
    NFC (near field communication) 28
    GPS 29
    Accelerometer, gyroscope, barometric altimeter, compass 29
    Heart rate sensor.......................................30
    Electrocardiogram (ECG).................................30
    Ambient light sensor....................................30
Tapping with Apple Watch's Haptic Feedback....................31

CHAPTER 2: **Time Out: Setting Up Your Apple Watch**............33
Setting Up Apple Watch 34
    Pairing Apple Watch with the iPhone 35
    Choosing a passcode for Apple Watch......................39
    Setting up cellular connectivity on Apple Watch.............41
    Monitoring the Apple Watch battery 43
Understanding the Home Screen 46
Maintaining Your Apple Watch 49
    Avoiding water for the original Apple Watch..................49
    Avoiding extreme temps 50
    Doubting its durability 50
    Considering a bumper 50
    Using Apple Watch responsibly...........................51
Taking Advantage of Accessibility Features....................54

CHAPTER 3: **Control Freak: Mastering Apple Watch's
Interface and Apps** 57
Handling Apple Watch's Controls 58
    Tap..58
    Double-tap 59
    Press (Force Touch).....................................59
    Two-finger press (time or heartbeat) 60
    Swipe..61
    Sketch 61
    Digital Crown button....................................62
    Side/Power button 63
Going Hands-Free with Siri...................................64
Vibrating Along with Apple Watch's Tactile Feedback 65
Using Control Center, Dock, and Notifications 66
    Control Center 66
    Dock...68
    Notifications 69

Looking at Apple Watch's Built-In Apps 70
   Phone 70
   Messages 70
   Mail 71
   Calendar 72
   Alarm. 72
   Stopwatch. 72
   Timer 72
   World Clock 72
   Siri 73
   Weather 73
   Stocks 74
   Activity. 74
   Workout 75
   Maps 75
   Photos 75
   Camera 76
   Music 76
   Remote 77
   Wallet and Apple Pay 77
   Apple News. 78
   Breathe 78
   Walkie-Talkie 78
   Heart Rate. 79
   Find Friends 79
   Podcasts 79
   Radio 80
   Audiobooks. 80
   App Store 81
   Noise 81
   Cycle Tracking 82
   Home. 82
   Calculator 82
   Voice Memos 82
   Settings 83
   Compass 83

PART 2: JUST THE TASKS, MA'AM! 85

CHAPTER 4: **It's About Time: Learning How to Set Watch Faces, Alarms, Timers, and More** 87
   Looking at the Built-In Watch Faces 88
   Activity (analog and digital) 88
   Astronomy 89
   Breathe 89

California 90
Chronograph 91
Color 92
Explorer 92
Fire and Water 92
Gradient 94
Infograph, Infograph Modular 94
Kaleidoscope 95
Liquid Metal 96
Mickey Mouse, Minnie Mouse 96
Modular/Modular Compact 97
Motion 98
Numerals Duo/Numerals Mono 98
Photos 100
Pride 101
Simple 101
Siri 101
Solar Dial 103
TimeLapse 103
Toy Story 104
Utility 104
Vapor 105
X-Large 105
Choosing from the Various Watch Faces 107
Differentiating Between Customizations and Complications 110
Customizations 111
Complications 111
Looking closer at a few complications .113
Accessing Time on Apple Watch 114
Accessing World Time 116
Taking Control: Alarms, Stopwatches, and Timers — Oh My! 117
Alarms 118
Stopwatches 118
Timers 121

CHAPTER 5: **Keep in Touch: Using Apple Watch for Calls, Messages, Texts, Walkie-Talkie, Emails, and More** 123
Accepting and Placing a Call on Apple Watch 124
Incoming calls 125
Outgoing calls 126
Outgoing calls with FaceTime audio 128

Handing Off a Call to Your iPhone or via Bluetooth. 130

Making Apple Watch Calls over Wi-Fi 131

Receiving and Sending Messages 133

Receiving and responding to messages. .133

Sending a message 139

Creating custom replies 141

Sharing your location from Apple Watch. 143

Deleting a conversation from Apple Watch. 143

Sending Stickers, Taps, Kisses, Heartbeats, and More. 144

Sketch 145

Tap. 146

Heartbeat 148

A few more Digital Touch options. 149

Stickers (including Animojis, Memojis), handwritten
messages, and Tapback 149

Enabling and Using the Walkie-Talkie Feature 152

Adding friends to Walkie-Talkie. 153

Starting a Walkie-Talkie conversation. 154

Sending, Receiving, and Managing Emails on Apple Watch 154

Reading and acting on an email 155

Replying to an email 156

Composing an email 158

CHAPTER 6: **In the Know: Staying Informed
with Apple Watch**. 161

Accessing Real-Time Weather, UV Index (UVI), and Wind Speed 162

Following Stock Information and Much More. 165

Adding Weather and Stocks to Your Watch Face Screen. 168

Using Dock on Apple Watch. 169

Launching Dock and more. 169

Customizing Dock 170

Mastering Notifications on Apple Watch 171

How notifications work. 171

Accessing the notification settings 172

Viewing notifications. 174

Changing how you get notifications on your Apple Watch 175

Accessing Your Calendar on Apple Watch. 176

Navigating the Calendar app 176

Responding to a calendar/appointment request 178

Accepting a calendar request through notifications 180

Setting Reminders on Apple Watch 181

Accessing Apple Watch's Integrated Calculator 182

Creating and Listening to Voice Memos. 183

Navigating the Maps App 184

Using the Compass 189

PART 3: IT'S ALL IN THE WRIST............................193

CHAPTER 7: **Siri Supersized: Gaining the Most from Your Personal Assistant**.........195

Setting Up Siri on Your Apple Watch 196
    Selecting a language 197
    Choosing a Siri voice and setting up search options 198
    Ready to rock? Connecting and talking to Siri................199
    Talking to Siri on Apple Watch..........................200
What Are Siri Shortcuts?..................................201
    Using the Siri watch face and Siri Shortcuts 201
    Using Siri effectively on Apple Watch......................202
    Clock/World Clock apps 202
    Messages app..204
    Phone/Contacts apps 204
    Mail app 205
    Calendar app 206
    Activity/Workout apps.................................206
    Maps app 206
    Music app 207
    Web searches..209
    Miscellaneous..209
Trying Other Tasks with Siri................................210
    Setting reminders by location 210
    Reading your texts....................................211
    Calculating numbers..................................211
    Finding your friends 212
Extending the Fun (and Silly) Ways to Interact with Siri.........212
    Say: "What's the best smartwatch?" 212
    Say: "I love you, Siri."................................213
    Say: "Siri, I'm bored."................................213
    Say: "Who's your daddy?" 213
    Say: "What's the meaning of life?" 213
    Say: "Will you marry me?".............................213

CHAPTER 8: **Fitness Fun and Happy Health: Apple Watch Is Your Workout Buddy and Digital Doctor** 215

Tracking Your Fitness with Apple Watch 216
Getting Up and Running with the Activity App 217
    Move 218
    Exercise..223
    Stand 224

Understanding the Workout App 225
Personalizing Reminders, Feedback, and Achievements 231
Reminders 232
Summary. 233
Achievements. 233
Using the Activity App on Your iPhone. 234
Checking Activity Trends. 236
Learning to Use (and Love) the Breathe App .238
Starting a Breathe session 238
Using the Apple Watch app to make Breathe changes 239
Cycle Tracking 240
Hearing Health: Using the Noise app 242
Advanced Health Help: Heart Rate, ECG, Fall Detection, and SOS. . . .243
How the heart rate sensor works 243
How the ECG monitor works 245
Setting up or editing your Medical ID information. 246
Checking heart rate. 247
Receiving heart rate notifications 248
Turning on Fall Detection. 249
What to do after a fall 250
Emergency SOS 251

CHAPTER 9: **Mucho Media: Managing Your Music,
Movies, and More**. 253
Using Apple Watch to Control Songs Stored on an iPhone. 254
Now Playing 256
Artists 256
Albums 257
Songs. 258
Playlists 258
Having Siri Play Your Music 259
Pairing a Bluetooth Device with Apple Watch. 260
Streaming Apple Music to Your Apple Watch 262
Playing Music from Your Apple Watch 262
Removing Music from Your Apple Watch 265
Playing Podcasts, Audiobooks, and Radio Plays. 266
Podcasts 266
Syncing podcasts to your Apple Watch 267
Audiobooks. 269
Radio plays 272
Controlling Apple TV and iTunes with Your Apple Watch 273
Setting up Home Sharing 273
Controlling Apple TV remotely. 274
Controlling iTunes remotely. 275

CHAPTER 10: **Pay for Play: Making Mobile Payments with Apple Watch**...................................277

Apple Watch and Security 278

Setting Up Apple Pay on iPhone 279

Setting Up Apple Pay on Apple Watch 281

Using Apple Pay with Your Apple Watch 282

Paying without a Nearby iPhone...........................284

Looking at the Wallet App 285

Using Apple Pay Cash on Apple Watch.....................288

Apple Pay Cash requirements.........................289

Setting up Apple Pay Cash in Wallet....................289

Using Apple Watch to send cash to friends................291

Using Apple Watch for Other Deals and Rewards 292

PART 4: MORE APPLE WATCH TIPS AND TRICKS 295

CHAPTER 11: **App It Up: Customizing Apple Watch with Awesome Apps and More** 297

Downloading Apps for Apple Watch........................298

From your iPhone 299

Directly to your Apple Watch..........................300

Deleting apps.....................................301

Adjusting Settings in the Companion Apple Watch App on Your iPhone 303

Settings and personalization.........................303

General options...................................304

Accessibility options 305

Security options...................................307

First-party app options 307

Twenty Recommended Third-Party Apple Watch Apps...........308

Mint 308

ESPN 309

Target 309

OneDrive..309

SPG 310

Chirp for Twitter...................................310

OpenTable 311

Evernote 311

American Airlines..................................311

BMW i Remote 311

CNN 312

eBay..312

Citymapper......................................312

TripAdvisor 313

NPR One 313

Fandango 314

Things 314

PayByPhone Parking 314

Sky Guide 314

Lutron Caséta 314

CHAPTER 12: **Extra! Extra! Having Fun with Apple Watch** 317

Copying Photos to Apple Watch 318

Launching Photos on Apple Watch 320

Discovering the Camera App 321

Examining a Batch of Apple Watch Games 324

Watch This Homerun 324

Trivia Crack 325

Runeblade 325

Best Fiends 327

Snappy Word 327

Watch Quest 328

Spy_Watch 328

Rules! 329

Wordie 330

Cosmos Rings 332

PART 5: THE PART OF TEN 333

CHAPTER 13: **Ten Cool Things to Do with Your Apple Watch** 335

Activity 336

Apple Pay 337

Hotel Key 338

Walkie-Talkie 338

Music Playback 340

Maps 340

Digital Touch 341

Siri 342

Watches and Watch Faces 344

Gaming 346

INDEX 349

Introduction

'm excited to present you with the third edition of *Apple Watch For Dummies* — your definitive guide to unlocking the power of your smartwatch.

In this book, you find out how to take full advantage of Apple Watch's many features, all in a language you can understand. You don't need a degree in electrical engineering to follow along with this book (badum bum!). Whether you're tech-shy or tech-savvy or perhaps somewhere in between, my goal is to teach you — in plain English — how to master your new gadget.

Apple Watch For Dummies covers all the things you can do with your sleek wrist-mounted companion, ranging from productivity and connectivity features to information, personalization, and navigation; health and fitness applications; entertainment options; and much more. And, of course, this book contains many visual examples of what you can do and what your watch should look like in certain situations, regardless of the model (or "Series") of watch you own.

About This Book

I wrote this book with one focus in mind: to cover all you need to know about Apple Watch in a language you can understand. I break down the geek-speak into street-speak.

After all, technology can be confusing, especially when it's a relatively new category of products like a "smartwatch." Therefore, consider this book your definitive guide to unlocking Apple Watch's capabilities. Kind of like with our brains, it was once believed we use only about 10 percent of what our consumer electronics products can do! — whether it's a computer, smartphone, tablet, TV, or camera — so it's my *modus operandi* to fill your head with the other 90 percent.

If you're more of an intermediate to advanced user, however, I also include a number of tips and tricks on how to get the most from your Apple Watch. Also, feel free to experiment — which can be part of the fun, but it may take some time to figure out all it can do. After all, you've probably used a computer for most of your life and a smartphone and/or tablet for the better part of a few years, but because a smartwatch is a bit of a different animal, it might take some trial and error — and this book — to master it!

You'll need to acclimate to the limited screen size, interacting with the watch via your fingertips and voice, and what apps work best on your wrist (just as you'll discover some things are simply better on a phone). Because Apple Watch might have a possibly steeper learning curve than other technology, give yourself a while to master it. Just know I'm here to help in a language you can understand, which means you can put away your geek-to-English dictionary.

Also keep in mind that Apple adds new features to Apple Watch over "time" or when a new operating system update is available. But don't fret: This book covers not only just the basics but also advanced capabilities. And hey, it's the third edition, after all and I'm keeping on top of what's new so you don't have to. And after you learn — and apply — a good number of the tasks discussed in this book, you should no doubt be comfortable with whatever new things the watch can do in the future.

On that note, Apple Watch is still in its infancy, so you're probably as excited as I am to see how this platform will mature. Think about how far the iPhone is today compared with the first generation in 2007. Ditto for the latest iPad when contrasted with the original in 2010. The best, as they say, is yet to come, but this book covers everything you need to know right now.

How to Use This Book

Although this book is meant to be a handy and informative resource, I hope you find the tone conversational. And as with other *For Dummies* books, you can read *Apple Watch For Dummies* in any order you like. I'd suggest you start with the first chapter or two in order to learn the various parts of the watch and its user interface, but after that, feel free to jump from chapter to chapter if one topic interests you more than another. Perhaps start by thumbing through the specific topics in the Table of Contents and then go to a particular chapter that piques your curiosity.

For example, you might wonder about the fitness capabilities of Apple Watch — how it counts your steps and the number of stairs climbed, calculates distance traveled, determines your calories burned and heart rate, and so on — so you can turn to Chapter 8 right away. Or maybe you're anxious to master text messages, emails, and calls on your new device? That would be Chapter 5 — on keeping in touch with those who matter. On the other hand, Chapter 10 focuses on using Apple Watch for making mobile payments at retail by waving your wrist over a contactless terminal to complete a transaction.

You get the idea. Each chapter can stand on its own.

While chapters are divided by task, be aware that each one also has a few subtopics within it. For example, Chapter 6 is on using Apple Watch to stay informed, but the information is broken down into such topics as calendar appointments, maps, live sports scores, stock quotes, and weather information.

But if you're "old school" and would like a more linear read, go ahead and flip through it from beginning to end — just don't expect a plot twist near the end. (Spoiler alert: The butler did it.)

In some cases, I cross-reference subjects with topics from other chapters whenever relevant, but you can skip over them if you like, or you can pursue them. I also cover how to best use your voice instead of your fingertips — after all, I wrote the book *Siri For Dummies* (shameless plug alert!).

It goes without saying that you'll benefit most from this book if you have your Apple Watch with you, along with your nearby iPhone — which may be required for some of the changes you'd like to make to Apple Watch — and if you ensure the battery is full on both devices so your lessons won't be interrupted. Oh, and it doesn't matter which Apple Watch you own — such as the original 2015 version or a Series 5, which adds an "always-on" screen, integrated compass, and new materials (such as ceramic) to choose from — because this book is relevant to every model that has come out over the years.

After trying many of the Apple Watch features I teach you in this book, expect to reach for your iPhone less and less. You'll likely hear this from seasoned Apple Watch users: Keep your iPhone tucked away in your pocket or purse but still access what you want by simply tapping or talking into your wrist. It's all about convenience.

The various tips and tricks throughout — as well as some interesting tidbits — will help you get the most from your Apple Watch. For example, you can use your Apple Watch to control music, audiobooks, and podcasts on your iPhone. Also, your watch knows the difference between a tap and a press. And did you know your watch can tap *you* with a slight vibration whenever a new message arrives for you to read? You learn how to do that — and much more — throughout this book.

But you don't need to wade through these extra Apple Watch factoids if you prefer to stick to the basics. Most of this extra content is labeled as Technical Stuff or Tip (see the "Icons Used in This Book" section). Then again, you might be more interested in these "sides" than the main course. (I'm sometimes like that when I visit my favorite restaurant.)

Also, be aware that this book has many figures, so you can see the steps — or the outcome of them — as you would on your own watch. This visual information should make it easier to follow along.

Foolish Assumptions

When writing this book, I made only two major assumptions:

>> You own one of the Apple Watch products.

>> You want to know how to get the most from it.

The watch won't come with an instruction manual, so consider *Apple Watch For Dummies* the closest thing to one — and a whole lot more too, if I may say so myself.

Icons Used in This Book

The following icons are placed in the margins of the book's pages to point out information you may or may not want to read.

TIP

This icon offers suggestions to enhance your experience. Most are tied to the topic at hand, whereas others are more general in nature.

REMEMBER

This icon reinforces the importance of information related to Apple Watch. You might consider bookmarking the page or jotting down the information elsewhere.

WARNING

Apple Watch is a promising new wearable platform, but this icon alerts you to important considerations when using it, including health, safety, or security concerns.

TECHNICAL STUFF

This icon warns you about geeky descriptions or explanations you may want to pass on — but don't expect a lot of these throughout this easy-to-read guide.

Beyond This Book

We're almost ready to dive into this book so you can master your Apple Watch. It should be noted that a lot of excellent online information tied to Apple Watch exists, so the following are a few good websites to check out and potentially bookmark for future visits:

- **Dummies** (www.dummies.com): The official website for all the Dummies books, including sections on Apple Watch, iPhone, and Siri.

- **Apple Watch For Dummies** **Cheat Sheet:** This site offers information discussed in this book but presented in different ways. To find the cheat sheet for this title, go to www.dummies.com and type "Apple Watch For Dummies cheat sheet." You'll find handy tips on using your watch to meet your fitness goals and getting turn-by-turn directions.

- **Apple Watch** (www.apple.com/watch): Apple's website for all things Apple Watch, including specifications, features, and an online store.

- **Wikipedia: Apple Watch** (http://wikipedia.org/wiki/Apple_Watch), a communally updated resource for Apple Watch, including links to other resources.

- **Gizmodo (Watch Section)** (http://gizmodo.com/tag/watch): A look at all watch-related information, photos, and reviews, including a ton of Apple Watch content.

- **9to5Mac** (http://9to5mac.com): A handy resource for all things Apple Watch, along with other Apple products.

- **Engadget** (www.engadget.com): Popular portal for consumer electronics, including coverage of Apple Watch news, reviews, and features.

- **The Loop** (www.loopinsight.com): Coverage of gadgets, apps, and other techy trends but with a strong focus on Apple products.

- **C|Net's Apple Watch page** (www.cnet.com/products/apple-watch): Apple Watch news, reviews, and features tied to Apple's first wearable.

Where to Go from Here

If you've never used an Apple Watch — perhaps you bought this book in anticipation of purchasing one or receiving it as a gift — it might be best to power up the watch, turn it on, and follow the prompts to set it all up. Chapter 2 goes into this if you prefer to wait, or feel free to dive in with the watch before you fully crack the spine of this book. Your call based on your comfort level.

Regardless of which model you own — or plan on buying or receiving — you don't need to know anything to begin reading *Apple Watch For Dummies*. All you need is your willingness to learn this exciting new wearable gadget, which should help add convenience, speed, and style to your everyday tasks.

Ready to start? Turn the page. . . .

1

Getting to Know Apple Watch

Learn how to set up your Apple Watch and discover its many features, including the Digital Crown button and the side button, and what aspects of Apple Watch make it unique from your iPhone.

Pair your Apple Watch with your iPhone and then learn about setting up a passcode and cellular connectivity, monitoring battery usage, and protecting your valuable investment.

Explore the many ways you can interact with your Apple Watch including tapping, pressing, and swiping as well as using Siri to help you complete watch-related tasks.

» **Understanding the different Apple Watch models**

» **Learning about the many features of Apple Watch**

» **Navigating the Home screen**

» **Exploring different parts of Apple Watch**

» **Understanding wireless capabilities and sensors**

Chapter **1**

Watch This: Introducing Apple Watch

S o, are you excited or what?

You're a proud owner of the trendy Apple Watch. Or perhaps you purchased this book in anticipation of picking one up or receiving it as a gift. Either way, thank you for reading *Apple Watch For Dummies*, 3rd Edition. This easy-to-read book has one goal in mind: to teach you everything you need to know about Apple Watch. With simple step-by-step instructions, clear images, and accessible tips and tricks, this book will help you gain the most from your new wearable gadget.

In this chapter, I walk you through the basics of Apple Watch to help you discover what this teeny wrist-mounted computer is capable of. You find out about the different parts of the watch — on the outside and inside — as well as the layout of the Home screen. From ways to interface with content on the watch to the hidden wireless technologies to integrated sensors that track your moves, you'll soon have a clear picture of the 21st-century magic you're wearing on your wrist.

Exploring the Different Apple Watch Collections

Apple Watch comes in a few different sizes and configurations. For Series 1, Series 2, and Series 3 watches, you have a choice of a screen that's either 38 millimeters (about 1.5 inches) or 42 millimeters (roughly 1.65 inches).

For Apple Watch Series 4 and Apple Watch Series 5, the screen sizes are a tad larger, but they don't look it; sizes measure 40 mm (1.57 inches) or 44 mm (1.73 inches) but because there are narrower "bezels" (borders) around the screen, the watch sizes themselves are virtually the same size as older options.

TECHNICAL STUFF

You measure the size of your screen from the top of the Apple Watch screen to the bottom and not diagonally — similar to how most screens from consumer electronics are measured (such as smartphones and tablets).

Although you've likely already bought a watch before buying this book, note that there are a few different versions of Apple Watch options available today — with the most popular version, Apple Watch Series 5, shown in Figure 1-1 — and a few accessories you can purchase to customize your watch. For a more extensive discussion about the Apple Watch collections, or for when you want to convince a friend or coworker that he or she needs an Apple Watch, visit `apple.com/applewatch`.

FIGURE 1-1: The most popular Series 5 Apple Watch has multiple band colors, styles, and material options too.

Excluding the various bands you can buy from Apple, the five different Apple Watch options are:

>> **Apple Watch Series 3:** Although not the newest Apple Watch available, this model was the first to offer either a GPS chip — to accurately capture location information — or a GPS and cellular option. More on this later in this chapter.

- » **Apple Watch Series 4:** This option was released in late 2018. Besides the slightly larger screen in the same-sized body, the series offers additional features including an ECG (electrocardiogram), fall detection, and more.

- » **Apple Watch Series 5**: At the time of writing this book, this is the latest Apple Watch, which debuted in the fall of 2019. Along with new band materials and styles, it added an "always on" screen, compass, emergency international calling, and more (which I cover in this book, of course).

- » **Apple Watch Nike+:** Ideal for fitness types who like the Nike brand, this special edition Apple Watch Series 4 (and special loop band) was designed to be your running partner and synchronizes with the Nike Run Club app and Nike Training Club app.

- » **Apple Watch Hermès:** A partnership between Apple and Hermès, this fashion-centric watch includes bold, colorful (and extra-long wraparound) leather bands and an exclusive new watch face.

You also have a ton of choice when it comes to material you want in an Apple Watch, and what style band to choose from. In fact, with Apple Watch Series 5, you can go with aluminum, stainless steel, titanium or ceramic! See Figure 1-2.

FIGURE 1-2: Wow! What a selection of different materials and bands to choose from with Apple Watch Series 5.

In the fall of 2019, Apple also announced Apple Watch Studio, a website that lets you choose a case and pair any band. Try it out for yourself at apple.com/shop/studio.

Figuring Out What Apple Watch Can Do

Some may question why they *need* a smartwatch. Perhaps you traded your watch for a smartphone years ago and now wonder why you'd go back to the wrist? One word: convenience. Not having to carrying anything is pretty darn handy, which you soon find out when using your Apple Watch. Simply glance at your wrist to glean information — wherever and whenever you need it. Not to mention your watch can *tap* you with a slight tactile vibration to let you know about something, such as a calendar appointment or a loved one giving you a virtual "poke." Buying something at a vending machine or a retail store by simply waving your wrist over a sensor is also kind of awesome. Or having an airline attendant scan a barcode on your watch's screen to let you board a plane? What a timesaver.

You can thus keep your iPhone tucked away, preserving its battery for when you really need to access something with it. In fact, some Apple Watch models can make or receive calls and texts even without a smartphone nearby, which I get to in Chapter 5.

Perhaps, because you wear it on your wrist and will likely glance at it multiple times throughout the day, Apple Watch will become an extension of yourself. When you strap it onto your wrist, you're not going to want to take it off. Now, that's personal.

As you discover in this book, Apple Watch has many, many features. Some of its main categories include time, communication, information, navigation, fitness, entertainment, and finance (mobile payments). The following sections highlight Apple Watch's main features, but be aware a few may require the GPS + Cellular model (I've indicated where).

Watch faces

Instead of a regular watch that simply shows one face, you can choose what you see on your Apple Watch. The watch has many styles to choose from right out of the box, as well as numerous downloadable apps that customize the look of the face. You can also change the color of the watch face to match your outfit. Chapter 4 walks you through it all.

Timers and alarms

Apple Watch also includes various stopwatches, timers, and alarms. Whether you use your fingertips or your voice, your Apple Watch can let you know when it's been 30 minutes so you can pull something from the oven. Or time your friend doing laps in a pool — from the comfort of your lounge chair. Apple Watch also lets you set an alarm to wake you up in the morning. You can use the Timer app as a game clock, for example, to tell you and your opponent when your time is up in a round of Scrabble. Check out Chapter 4 for all the details.

Caller ID, or even calls

See who's calling by glancing at your wrist. Apple Watch displays the caller's name (Caller ID) or perhaps just a phone number (which often happens if that person isn't in your iPhone's Contacts). You can also use the Apple Watch microphone to record and send sound clips to friends. Some Apple Watch models — those advertised as GPS + Cellular — let you leave your iPhone at home and take or make calls right from your wrist when you're out! Heed the call and go to Chapter 5 for details.

Walkie-Talkie

What's more fun and quicker than a phone call? The Walkie-Talkie feature built into Apple Watch. As the name suggests, Walkie-Talkie lets you press to talk to someone else who has an Apple Watch. Let go to listen for the reply. 10-4, good buddy? I cover this in Chapter 5, which is about the different ways to use Apple Watch to communicate.

Emergency SOS

In a nutshell, the Emergency SOS feature built into Apple Watch calls for help — when you can't. Whether it senses a troubling anomaly through the heart-rate or ECG sensors, or detects a fall, Apple Watch can dial 9-1-1, notify your emergency contacts, send your current location, and even display your Medical ID badge for emergency personnel. I cover all this in Chapter 8, which focuses on fitness and health.

Text messages

You can read and reply to text messages with Apple Watch, as shown in Figure 1-3. Hold your wrist up to read the message or lower your arm to dismiss it. Chapter 5 walks you through all the text messaging functions for Apple Watch, including some models that don't require your iPhone to be near you at all!

FIGURE 1-3:
Read and reply to messages on your Apple Watch.

Email

When an email comes in, you can read it on your wrist (scroll up and down the screen with your fingertip to see all the text), flag it as something to reply to later, mark it as read (or unread), or move it to the Trash. As with text messages and phone calls, you can transfer email from Apple Watch to your iPhone to pick up where you left off. I cover all this in Chapter 5.

Wrist-to-wrist communication

Along with the Walkie-Talkie feature, your smartwatch lets you communicate directly with someone else's wrist via a component called *Digital Touch*. For a sketch, use your fingertip to draw something, such as the heart shown in Figure 1-4, and the person who receives it will see it animate — just as you drew it. Or why not send some virtual kisses to let someone know you're thinking about them. As described in Chapter 5, you can even send your heartbeat to someone by pressing two fingers on the screen.

Dock

Naturally, a wearable watch is a convenient way to stay on top of important information. Apple Watch has a cool feature called *Dock* that lets you quickly open your favorite apps or go from one app to another. To launch the Dock, press the side button and swipe up or down (or turn the Digital Crown). See Figure 1-5. Chapters 3 and 6 discuss how to access Dock, customize what you see, and scroll through relevant information.

Calendars

Apple Watch also has a Calendar app (with reminders) so you can stay on top of events throughout your day (or coming in the near future). Also, when you receive a calendar invitation, you can immediately accept or decline it on your wrist and even email preset responses to the organizer. Put Chapter 6 on your calendar for more information.

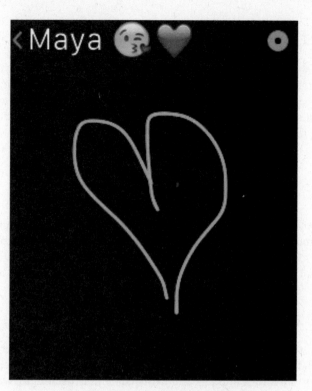

FIGURE 1-4:
Sketch
something on
your Apple
Watch and
send it off to
someone else's
Apple Watch.

FIGURE 1-5:
Dock allows
you to quickly
open your
favorite apps
or jump from
one app to
another.

Maps

Your wrist is an ideal place to glance at a map. Get turn-by-turn directions from your current location — and you don't have to worry about having to stare at your wrist for visual cues (or fall down a manhole in the process) because Apple Watch gives you a tap on the wrist to let you know when it's time to turn left or right. Navigate to Chapter 6 for more. With Apple Watch Series 5, an integrated compass helps you navigate even further (unleash your inner Boy Scout or Girl Guide)!

Siri

Just as you can talk into your phone, Apple Watch also has a microphone, which means you can have access to your personal assistant known as Siri. Flip to Chapter 7 to find out more about what Siri can do for you. As the author of *Siri For Dummies*, I share some of my favorite Siri tips and tricks you can master with ease.

Fitness

One of the coolest applications for Apple Watch? Fitness. Chapter 8 looks at using the watch to measure your activity — steps, stairs, distance, time, calories burned, and heart rate information — and to display it in a meaningful way on your watch and smartphone. I cover the Activity app, shown in Figure 1-6, and its three rings, which show you relevant information on your daily activity (or lack thereof!). On the other hand, the Workout app (as shown in Figure 1-7) offers some workout routine options — including walking, jogging, running, and cycling — and shows real-time stats on your cardio session.

Music connectivity and more

Chapter 9 teaches you how to use Apple Watch like a wireless remote. Control your music on your phone — from the convenience of your wrist — as well as listen to synced playlists on your watch *without* needing your iPhone (but with Bluetooth headphones). Along with talking about music streaming and downloads, I highlight how to manage podcasts, audiobooks, radio plays, and other audio. Chapter 9 also covers how to control Apple TV on your Apple Watch.

Apple Pay

Swiping your wrist at retail stores or at a vending machine is super cool, and Chapter 10 covers all the ways you can use your Apple Watch in this regard. Your watch lets you buy products and services via Apple Pay — and you don't even need your iPhone with you.

FIGURE 1-6:
The Activity app shows three rings that summarize your daily progress — so far.

FIGURE 1-7:
The Workout app offers you some different exercise routines to choose from.

Other apps

Apple Watch is quite a versatile gadget, which means other apps can help enhance its convenience. Chapter 11 looks at a number of optional third-party apps you can download to further personalize the most personal gadget in the world.

Other functions

Chapter 12 takes a closer look at some of the extra fun things you can do with Apple Watch. I cover using your wrist to remotely snap a photo on your iPhone, as well as look at photos on your wrist, including how to zoom into a photo (because maybe you're bored in line at the supermarket and want to see some smiling faces or furry pets). I also discuss Apple Watch as a gaming platform and what's available.

Bonus tips

Chapter 13 reveals the top ten things you should try with Apple Watch and, of course, how to pull them off with grace. I share the absolute coolest things this smartwatch can do and how to best demonstrate them to your friends — to the point they'll be boiling with envy.

TECHNICAL STUFF

Apple says you could squeeze up to 72 hours of battery life on Apple Watch, but be aware this varies greatly on how often you use the watch, the settings you choose (see Chapter 2), what apps you use, outside temperature, and other factors. Apple says most people will find it lasts up to 18 hours — hence, the "all day." The test involved 90 time checks, 90 notifications, 45 minutes of app use, and a 60-minute workout with music playback from Apple Watch via Bluetooth.

Determining What You Need for Your Apple Watch

The original Apple Watch didn't do too much on its own. Rather, think of it as more of a companion device to an iPhone. Oh, sure, it could do a few things by itself — such as show you the time, count your steps, make payments, and play music — but a wirelessly tethered iPhone was required for the overwhelming majority of features.

But now with Apple Watch Series 3, 4, and 5, you can go with one of two models:

>> **GPS:** This is great for navigation.

>> **GPS + Cellular:** With this, you can pay your mobile phone provider to unlock the eSIM (a virtual SIM card inside Apple Watch) so you can use the Apple Watch on the go like a phone. It can take calls, text messages, stream music, and more! In North America, this service costs $10/month because it's added to your existing smartphone plan.

This book is ideal for whichever model you have, so not to worry. As mentioned before, you do need an iPhone to set up Apple Watch, even if you have the version that doesn't require one nearby to work.

If you do own an older model (Series 1 or Series 2), you need at least an iPhone 5 to use Apple Watch. For those who own a Series 3, Series 4, or Series 5 model, you will need an iPhone 6 or newer. You also need to download and install the latest iOS operating system update from Apple — whether you do it on your iPhone or on iTunes (on a PC or Mac) — and then connect the iPhone to your computer with a USB cable. After you download the latest operating system, an Apple Watch app — a white watch against a black background — appears on your iPhone's Home screen, as shown in Figure 1-8.

In fact, you can now install the latest WatchOS update "pver the air" (OTA) without using a physical connection at all!

The Apple Watch app

FIGURE 1-8: Whether you own an Apple Watch or not, an Apple Watch app (shown at top right) appears on your iPhone's Home screen.

Getting to Know the Apple Watch Home Screen

As shown in Figure 1-9, the main Home screen of Apple Watch is populated by a number of small bubble-like icons. It's quite neat actually, not to mention functional. Simply tap an icon with your fingertip to open an app, or slide around the Home screen to see other icons pop up and grow larger (you want the app centered on the screen for easy access).

FIGURE 1-9: Press and move your finger around to see all the apps on your Home screen (or twist the Digital Crown button to zoom in and out).

If you're an iPhone user, the icons should be familiar to you; therefore, you know what built-in and third-party apps launch when you tap a specific icon. Table 1-1 shows some of the built-in apps. See Chapter 3 for more on the native Apple Watch apps.

TABLE 1-1

Built-In Apple Watch Apps

App	Icon
Activity	
Alarm	
App Store	
Apple Store	
Audiobooks	
Breathe	
Calculator	
Calendar	
Camera	
Clock	
Compass	

(continued)

TABLE 1-1 *(continued)*

App	Icon
Cycle Tracking	
Find Friends	
Heart Rate	
Home	
Mail	
Maps	
Messages	
News	
Noise	
Now Playing	
Phone	

App	Icon
Photos	
Podcasts	
Radio	
Reminders	
Remote	
Settings	
Stopwatch	
Walkie-Talkie	
Wallet & Apple Pay	
Weather	
Workout	
Voice Memos	

You no longer need a nearby iPhone to install new Apple Watch apps — so long as you're running the WatchOS 6 operating system or later. (You can tell which version you're running by going on the Apple Watch to Settings ⇨ General ⇨ About.) In Chapter 11, I cover both ways to install new Apple Watch apps, whether you want to still use an iPhone (or Mac/PC), via the Apple Watch Store, or download directly to the device.

And, of course, third-party apps have their familiar icons, such as a big P for Pinterest, a "swoosh" for Nike, a green leaf for the Mint app, and so on. You can bet many thousands of developers will have their apps working on Apple Watch right out of the gate, but be patient in case some of your favorites aren't ready just yet. And some just might not make sense on a watch, such as an epic fantasy role-playing game that works better with a larger screen.

Learning About Apple Watch's Parts

Okay, so you're all geared up to test drive all that Apple Watch can do, but if you're using it for the first time, you may not even know all the parts of the watch and what they do.

Fair enough. In this section, you discover the basics of the hardware itself. I start with a look at the various parts of the watch on the outside and what they do.

Watch face

Regardless of which size you opted to go with — 38 mm, 40 mm, or the 42 mm model (referring to its vertical height) — the Apple Watch face is entirely digital; therefore, you won't find any buttons of any kind. Use your fingertip to move around the icon bubbles and tap an app to launch it. You can also tap, press, and swipe inside an app to perform a task.

You don't need to press hard on these buttons or on the watch face. You want to minimize the wear and tear of your new (and pricey!) gadget. Just a simple press on the buttons and watch face will do. And although Apple Watch Series 2 and newer are waterproof, try to avoid touching the screen and buttons with wet or damp hands. (Apple says "we recommend not exposing Apple Watch to soaps, shampoos, conditioners, lotions, and perfumes as they can negatively affect water seals and acoustic membranes.") See Chapter 3 for more on these buttons and using your fingers with your Apple Watch.

Digital Crown button

Seasoned watch owners are familiar with the small rotary dial on a watch's right side (left-handed people may flip the watch around so it's on the left side), which is used to either wind it up (for the old-school ones anyway) or set the time. Apple Watch has one too. Called the *Digital Crown button* — shown in Figure 1-10 — this dial can be pressed, tapped, or turned forward or backward, with each change resulting in a different action. See Chapter 3 for more on what the Digital Crown button can do.

Digital Crown

FIGURE 1-10: If worn on the left wrist, the side button is on the right-hand side of the watch case. The Digital Crown button is the ridged dial.

side button

Side button

Along the side of the watch is a long button, called "side button" (how imaginative!), as shown in Figure 1-10. From the Home screen and in any app, press this button to pull up your Dock (more on this later). Press and hold the side button to use SOS; double-click to use Apple Pay; or press and hold to turn your Apple Watch on or off.

Back sensors/ charger

On the back of Apple Watch, as shown in Figure 1-11, are multiple sensors to monitor your heart rate (measured in beats per minute) and more. A larger circle surrounding the four smaller heart rate sensors is used to charge up the Apple Watch battery. See Chapter 2 for more on the sensors and the charger.

Watch band

Every wristwatch has a band to keep the screen snug on your wrist. You chose a specific band when you bought Apple Watch — whether a leather strap, a link bracelet, a classic buckle, a silicone band, or other materials and styles — but you can change bands later if you desire. Apple introduced a slew of new bands in late 2019, too.

FIGURE 1-11:
The sensors on the back of Apple Watch can, among other things, calculate your heart rate.

Using Apple Watch's Touchscreen

Just like you can interface with a smartphone, tablet, and laptop in different ways — based on the task at hand — Apple Watch gives you three ways to use the small screen on your wrist:

>> **Tap:** Tapping with one finger on Apple Watch performs the same function as you'd expect on a smartphone: It selects whatever you're tapping, such as an icon to launch an app, a song to play a track, a link to a website, a photo to enlarge, or virtual buttons, such as on a calculator. On the Home screen, you tap and slide your finger around to move the icon bubbles. A tap is like a left-mouse click on a computer.

>> **Press:** Apple Watch knows the difference between a quick tap and a longer press — usually when you need to open some additional menus. Think of it as a kind of right-mouse click. For example, tapping a song plays the track, but pressing and holding it opens a set of options: Shuffle, Repeat, Source, and AirPlay. The technology that senses the difference between a tap and a press is called *Force Touch*.

>> **Swipe:** Many of the areas of Apple Watch — like Dock — and most of the apps you can access let you swipe left and right or up and down to navigate between different screens. For example, in Workout mode, you can see time elapsed as well as heart rate info, but swipe to the side to pull up music that you can pause and play. Swipe one more time and you'll see some options including the ability to lock your watch so you don't accidentally tap the screen during rigorous exercise, to pause your counter, and so on.

TIP

Some features are activated with two fingers pressed on the screen. In Chapter 5, you can find out how to record and send your heart rate or heartbeat to a loved one's Apple Watch.

Understanding Apple Watch's Wireless Functions and Internal Sensors

Oh, Apple Watch, you cleverly hide so much of your magic under your skin.

While Apple isn't revealing much about the "brains" of Apple Watch — a system-on-a-chip called the $S1$ — a more interesting dissection is provided by all the other goodies that make Apple Watch a fully functional wearable.

Apple Watch indeed houses a good number of wireless radios beneath its surface, including Bluetooth, Wi-Fi, GPS, NFC, and more. To better understand what they do, consider the following sections.

Bluetooth 4.0

Bluetooth makes a local wireless connection between two or more devices. Just as your wireless headset is paired with your smartphone so you can make hands-free calls, Apple Watch wirelessly communicates with a nearby iPhone. This lets you see texts on your watch, receive phone calls, control your music on your phone, and more. Bluetooth 4.0 works with devices up to 200 feet away (compared to just 30 feet with earlier versions). If you have an Apple Watch that supports cellular connectivity and pay for the service, you can perform many of these features — calls, texts, and accessing online music — without a nearby iPhone.

Wi-Fi

Even if you don't have a cellular model, Apple Watch features Wi-Fi, which gives it online connectivity — even when no iPhone is in sight. As long as you're on a wireless network, such as your home's Internet connection or a coffee shop's hotspot, you can access such information as email, live sports scores, mapping information, and so on. A feature called *Continuity* — introduced in iOS 8 — means you can also receive messages and take calls on multiple iOS devices (such as answering a call on your iPad) as long as you're in range of your Wi-Fi network; and Apple Watch can do this, too. See Chapter 5 for how to take advantage of Bluetooth and Wi-Fi connectivity.

Cellular

As I mention previously in this chapter, Apple Watch Series 3 and Series 4 have models called GPS and GPS + Cellular. As you can guess, the GPS + Cellular watches cost a little more, but it means you can make a call, send a text, and stream Apple Music from your wrist — all without your iPhone. You will need to pay your mobile phone provider an extra amount per month (usually $10/month) to activate the eSIM inside Apple Watch. That is, you don't need to insert a physical SIM card, like the one in your iPhone, to access LTE and UMTS cellular bands. That said, be aware there is no Apple Watch model that supports worldwide roaming.

For business travelers, Apple watch models can be purchased for specific locales: Europe/Asia Pacific and China mainland, for example.

With Apple Watch Series 5, emergency calling now works in countries outside of the U.S. (see Chapter 5).

NFC (near field communication)

NFC is a short-range radio technology (like Bluetooth) that has a number of applications but is most commonly associated with mobile payments. Similar to waving or tapping your iPhone 6 or iPhone 6 Plus on a contactless terminal at retail locations (or a compatible vending machine) to make a secure purchase, Apple Watch also uses NFC to make a *digital handshake* with the terminal to complete the transaction. Yep, it's all in the wrist. This is part of Apple Pay, Apple's mobile payment solution for secure cash- and card-less payments. Check out Chapter 10 for more on Apple Pay.

GPS

Except for the first Apple Watch (Series 1; from 2015), Apple Watch has an integrated GPS chip to identify its location on Earth down to a few meters of accuracy. Therefore, when coupled with mapping applications, GPS can help you see your location on a map, get directions from point A to point B, look for local businesses of interest, and more. GPS can also help with tracking fitness data when measuring steps won't help (such as in cycling). Along with the accelerometer (discussed next), built-in heart rate sensor, and Wi-Fi, Apple Watch's GPS can help measure distance traveled. Jog on over to Chapter 8 to learn more about the Activity and Workout apps.

TIP

Ever want to take a screenshot of something on your Apple Watch, such as an impressive day of physical activity? Press and hold the side button and then tap the Digital Crown. You'll hear a shutter button, the watch face will flash white, and the image will appear wireless in your iPhone's photo gallery. You may need to first enable screenshots by opening the Apple Watch app on an iPhone, and then tapping My Watch ⇨ General ⇨ Enable Screenshots.

Accelerometer, gyroscope, barometric altimeter, compass

As with other smartwatches and activity bands on the market, Apple Watch has an accelerometer that measures movement — whether you're lifting the watch to your face to turn on the screen, lowering your wrist to not accept a call, or calculating fitness activities, including your steps taken (like a 21st-century pedometer), total distance traveled, time spent exercising, and estimated calories burned. Beginning with Apple Watch Series 3 (2017), the watch also has an integrated barometric altimeter for measuring elevation — like counting the steps you climb or descend — as well as calculating altitude, for those who ski, hike, or mountain climb!

With the Apple Watch's accelerometer and gyroscope sensor, Apple Watch Series 4 (2018) and Apple Warch series 5 (2019) can also detect if you've fallen, and you can initiate a call to emergency services (or dismiss the alert). If you're unresponsive after 60 seconds, Apple Watch automatically places the emergency call and sends a message with your location to your emergency contacts.

A compass has been added to Apple Watch Series 5, which always points you north inside of apps like Maps, plus there's a dedicated Compass app, too. (Cue the song "Go West" by the Pet Shop Boys or the Village People before that!)

Heart rate sensor

A custom heart rate sensor included with Apple Watch helps you in two ways.

>> **Gauging your exercise intensity and tracking overall calorie burn:** (This is an estimation based on info you first input, required only once, such as your height, weight, gender, and age.) Apple Watch listens to your heart's beats per minute (BPM) and shows you data on the screen — if and when you call for it.

>> **Tracking your heart rate throughout the day:** Apple Watch can alert you if it detects unusually high or low heart rates — and yes, you can set the parameters if you want, even if you don't feel symptoms. Behind the watch are multiple sensors that measure your pulse through your skin. Going beyond fitness are the fun applications too, such as sharing your heartbeat with someone — felt on his or her Apple Watch — to show you're thinking about that person. See Chapter 5 for how to share this information to your heart's content.

Electrocardiogram (ECG)

Apple Watch Series 4 and Apple Watch Series 5 models include an electrical heart rate sensor that can take an electrocardiogram (ECG) using an ECG app; the built-in sensor and the electrodes are included in the Digital Crown button. According to reports, this app may only be available in the U.S. Once it does become available, you would take an ECG reading by placing a finger on the Digital Crown while wearing Apple Watch to have the reading completed within just 30 seconds. The ECG app tracks whether your heart is beating in a regular pattern or if you have signs of atrial fibrillation, a clear indicator of serious health problems.

Ambient light sensor

Finally, Apple Watch has an ambient light sensor under the glass, which samples the environmental light falling on the screen and then automatically adjusts the brightness up or down to improve viewing comfort (so it's not too dim or too bright, based on where you are). An ambient light sensor also helps to regulate the power that the display uses, thus squeezing more battery life out of the watch.

In fact, with Apple Watch Series 5, the Retina screen is always on — just dimmed — until you turn it towards your face to read the time or other info. In previous Apple Watch models, the screen goes to black when you're not looking at it, and takes a split second to wake up and turn on when you turn it towards you.

Tapping with Apple Watch's Haptic Feedback

You can tap Apple Watch's screen, but guess what? It can tap you, too.

As with video game controllers that vibrate (when your soldier gets shot) or some smartphones and tablets (that *buzz* slightly when you tap a letter on the virtual keyboard), Apple Watch employs *haptic technology* to apply light force to your skin about relevant information.

Apple calls it *Taptic Engine,* a linear actuator inside the watch that produces discreet haptic feedback.

Consider this slight vibration a third sense (touch), along with sight and sound, to give you information. The physical sensation of a tap tells you something, such as a warning that an important meeting is about to start without you even having to look down at your wrist. It can also be a silent alarm clock to wake you up in the morning instead of bothering your significant other. Or it can transmit the feeling of your loved one's heartbeat even though he or she may be miles way.

What's more, Apple Watch can tap different patterns based on who's reaching out to you (such as two taps for your spouse and three taps for your boss), or perhaps the haptic pattern tells you what the information is (one tap for the time on the hour, four taps for a calendar appointment, and so on). Neat, huh?

In the near future, Apple Watch's haptic feedback may let you know about important health information — perhaps when working in conjunction with sensors. Imagine if someone living with diabetes could feel a haptic tap on the skin to tell him or her it's time to take insulin based on the body's blood sugar levels.

IN THIS CHAPTER

» **Setting up the watch for first-time use**

» **Choosing a PIN**

» **Setting up Cellular on Apple Watch (on select models)**

» **Charging Apple Watch**

» **Taking care of your smartwatch**

» **Using Apple Watch responsibly**

» **Taking advantage of accessibility features**

Chapter **2**

Time Out: Setting Up Your Apple Watch

Well, you did it. You're now the proud owner of an Apple Watch.

Or perhaps you received one as a gift and you're more intimidated than proud?

Regardless, I'm thrilled you picked up (or downloaded) this book to help you get the most from your wearable companion. You're gonna love your new gadget.

This chapter helps you set up Apple Watch for the first time and covers how to charge it, take care of it, and get to know some of the basics.

Setting Up Apple Watch

If you're a fan of those unboxing videos on YouTube, you've watched a few gadget geeks excitedly open a box for the first time and expose all the goodies inside. But if you don't have the time or desire for that drama, let me tell you what you can expect if you haven't received your Apple Watch just yet or what you should find in your box if you have.

In your Apple Watch box, you should find:

» Apple Watch

» Band (what kind will vary on the model)

» Magnetic dock charger cable (USB cable)

» Five-watt power adaptor (white plug for the wall)

» Small booklet with setup and maintenance tips

Be sure to keep the box and all the things inside just in case you need to return or exchange the watch.

Like many consumer electronics you buy today, the watch might be already charged when you first get it, but it's always a good idea to plug it into a computer or the wall to give it a full boost before using it for the first time. This way, you won't run out of juice while playing around with the watch's settings, learning the mechanics, and so on.

DID YOU KNOW?

Beginning with Apple Watch Series 2, you can wear all models in the shower, outside in the rain, while sweating during a workout, or while swimming! But be aware, Apple Watch is meant for "shallow water" activities and you should not wear it scuba diving, water skiing, or doing "other activities involving high velocity water or submersion below shallow depth," cautions Apple. Also, avoid soap and shampoo if you can because this could impact the watch's performance. Near the end of this chapter I share some obvious (and not so obvious) ways to take better care of your Apple Watch.

Pairing Apple Watch with the iPhone

After you ensure your smartwatch is charged, follow these steps:

1. **Turn on your Apple Watch by pressing and holding the side button.**

This is the one flush with the watch and not the Digital Crown that's sticking out. You will see the Apple logo appear in the center of the screen. This is a good sign!

2. **Tap the Apple Watch app on your iPhone.**

It's a black icon that simply says Watch. If you don't see the app on your phone's Home screen, swipe left or right to look for it.

TECHNICAL STUFF

You need an iPhone 5 or newer and the 8.2 iOS (or newer) operating system installed to use Apple Watch for Series 1 and 2, and an iPhone 6 and iOS 12 or later for Series 3, Series 4, and Series 5. To double-check what you have, tap Settings ➪ General ➪ About and look where it says Version. Your phone also notifies you about any available updates.

Once you bring the Apple Watch close to your iPhone, you should see the words "Use your iPhone to set up this Apple Watch" appear on your iPhone screen. If you don't, open the Apple Watch app on the iPhone and then tap Start Pairing.

3. **Tap Continue.**

Now, keep your Apple Watch and iPhone close together until you complete these next steps.

4. **Follow the prompts.**

They ask you to hold Apple Watch up to the iPhone's camera. You can then align the watch's face within the viewfinder in the center of the screen. This should do the trick. See Figures 2-1 and 2-2 for a look at setting up Apple Watch for the first time.

If that doesn't work, tap the Pair Apple Watch Manually option, in orange, at the bottom of the app. You're prompted to tap the "i" ("information") app on your Apple Watch to view its name and then tap the corresponding name listed in the app. If it's not listed, be sure your wireless connection is enabled; then, swipe up from the bottom of the screen and tap the icons for Bluetooth and Wi-Fi so they're highlighted and not grayed out.

TIP

If you're prompted to set up cellular connection, read the section "Setting up cellular connectivity on Apple Watch" further on.

We're almost done!

5. **If this is your first Apple Watch, tap Set Up Apple Watch.**

Otherwise, choose a backup. If prompted, update your Apple Watch to the latest version of watchOS, the name of Apple's operating system that powers this wearable.

6. **Read the Terms and Conditions and tap Agree (twice).**

If asked, enter your Apple ID password. If you aren't asked, you can sign in later from the Apple Watch app (General ⇨ Apple ID). Some features that require a cellular phone number won't work on cellular models of Apple Watch unless you sign in to iCloud.

FIGURE 2-1:
The Apple Watch app asks you to pair your Apple Watch. Easy peasy.

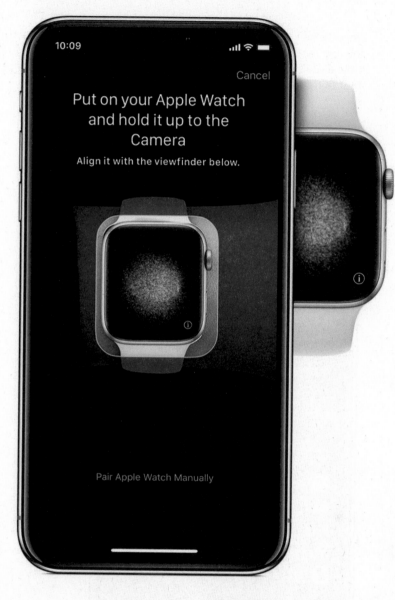

FIGURE 2-2:
Match up the
Apple Watch
inside the
outline on your
iPhone screen.

Your Apple Watch will show you which settings it shares with your iPhone. For example, if you've enabled Find my iPhone, Location Services, Wi-Fi Calling, and Diagnostics on your iPhone, these settings automatically turn on for your Apple Watch. You can select other settings, too, like Siri (your personal voice-activated assistant) and Route Tracking.

Once the pairing is successful, you can adjust additional watch settings from within the app by tapping My Watch at the bottom left of the screen (see Figure 2-3). Take some time to familiarize yourself with this great app — I spend a lot of time on it in this book too.

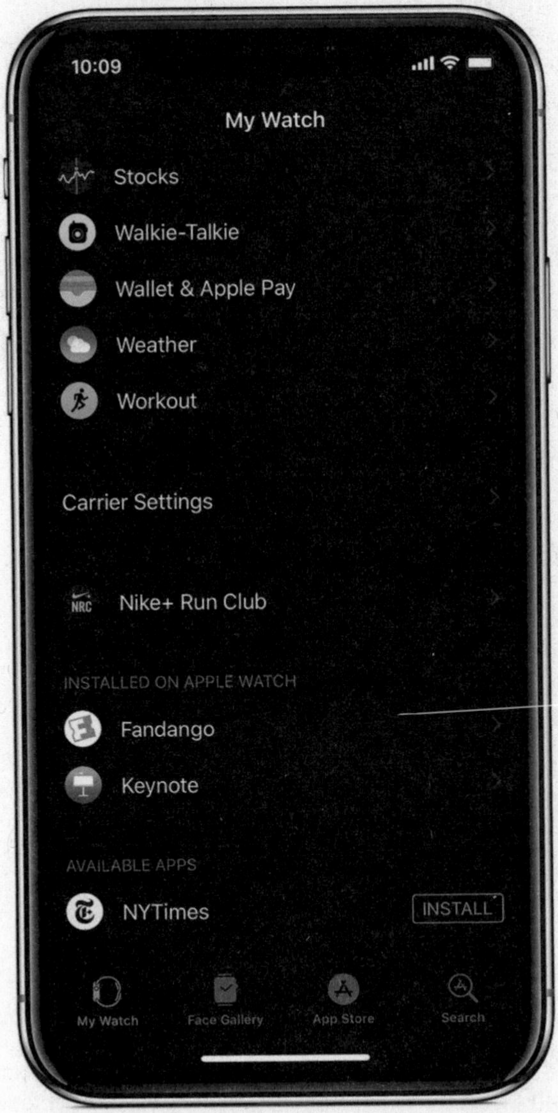

FIGURE 2-3: While you might be anxious to play around with your new Apple Watch, spend some time familiarizing yourself with the Apple Watch app on iPhone as well.

- » **Face Gallery:** Tap this to view and change your Apple Watch clock faces (see Chapter 4 for all your options).

- » **App Store:** This takes you to a place to download and install third-party apps to your device.

- » **Search:** Tap this to search for something by keyword or phrase.

Keep in mind that you don't need to turn Apple Watch on or off. Simply raise your wrist and the screen turns on — thanks to its internal accelerometer (motion sensor) — and lower your arm to turn it off. It's that easy. Or with the latest Apple Watch Series 5, the screen never goes dark — it stays on but dims itself until you look at it!

TECHNICAL STUFF

How does Apple Watch Series 5's beautiful Retina display stay on all the time without killing the battery? At the risk of geeking out, it uses a LTPO display — a "low temperature poly-silicon and oxide" display — that drops the screen's refresh rate from 60 Hz down to a "power-sipping" 1 Hz when the watch is inactive (that is, when you're not looking at it). A new low-power driver, ambient light sensor and efficient power management software also work together to keep your watch going up to 18 hours between charges. Just touch the screen or point it towards your face for full brightness. Cool, eh? See Figure 2-4.

Choosing a passcode for Apple Watch

Although not mandatory, those who use Apple Watch can set up a personal identification number (PIN) in case the watch is lost or stolen. Some people call it a *passcode* instead of a PIN, but it's the same thing.

TECHNICAL STUFF

A PIN is mandatory when you use Apple Pay to buy goods or services through your smartwatch. See Chapter 10 for more on Apple Pay.

Just like with an iPhone, iPad, or iPod touch, if the Apple Watch's PIN option is activated, you have to enter the four-digit numeric code (or you can choose a longer one) every time you put the watch back on. Apple Pay requires skin contact and the PIN. See Figure 2-5 for setting up a PIN on Apple Watch.

Wait, how does Apple Watch "know" if someone is taking the watch off or putting it on? It's because of the smartwatch's internal sensors that touch your skin. I cover this more in Chapter 10, where I also go over what to do if you lose your Apple Watch. (Hint: As with a lost iPhone, you can go to www.icloud.com, log in with your details, and remove your credit and debit cards in case the watch falls into the wrong hands.)

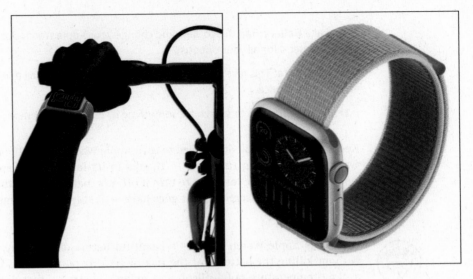

Bottom line: You may want to set up a passcode on your Apple Watch even if you don't activate Apple Pay. And, of course, you should also have a PIN set up on your iPhone.

After you set up a passcode, you may need to wait a little for your devices to sync. It really depends on how much data you have, but should only take a few minutes — even at the longest. Be sure to keep your devices close together until you hear a chime and feel a gentle tap from your Apple Watch, to confirm they have synched together. Now press the Digital Crown to see your apps.

FIGURE 2-5:
A passcode is optional for Apple Watch unless you use Apple Pay.

Setting up cellular connectivity on Apple Watch

If you're reading this section, then you've got one of the Apple Watch models that supports cellular connectivity. This is super exciting because you can use your watch for calls, texts, real-time notifications, streaming music, and more — even when you're away from your iPhone.

That is, non-cellular models require Apple Watch to be near an iPhone for most "connectivity" tasks, like calls, messages, accessing online content, and receiving notifications.

To use your Apple Watch's cellular functions, you will need:

» Apple Watch Series 3, Series 4, or Series 5 (GPS + Cellular)

» iPhone 6 or later with the latest version of iOS

» An eligible cellular service plan with a supported carrier (iPhone and Apple Watch must use the same carrier)

Setting up your GPS + Cellular watch is easy. You can follow along with the onscreen steps (shown in Figure 2-6). Alternatively, you can set up cellular later from the Apple Watch app. Using the Apple Watch, the following steps walk you through connection and use of the watch afterwards:

1. **When you first set up your new Apple Watch, follow the prompts to activate cellular (you can see the start screen in the left image of Figure 2-6).**

   Once activated, note the color of the Cellular button on your Apple Watch:

   - *Green:* Indicates you're using cellular connectivity, and the green dots over the button show the signal strength (the more dots, the stronger and better the signal). You can see this in the middle image of Figure 2-6.

   - *White:* As the right image in Figure 2-6 shows, the once-green button turns white when your cellular plan is active, but your watch is connected to your iPhone using Bluetooth or Wi-Fi.

FIGURE 2-6: The Apple Watch app on iPhone walks you through the setup process.

2. **On your iPhone, open the Apple Watch app, tap the My Watch tab in the lower left of the screen, then tap the green icon that says Cellular.**

3. **Follow the instructions.**

Now, going forward, your Apple Watch automatically reverts to the most power-efficient wireless available, whether it's a nearby iPhone via Bluetooth, a Wi-Fi network, or cellular connectivity. When your watch connects to cellular, it uses LTE (4G or "Long Term Evolution" networks). If LTE isn't available, your watch attempts to connect to UMTS (should your carrier support it).

Some tips and tricks to remember:

>> **To check the signal strength of your carrier network:** Swipe up from the bottom of the Apple Watch screen to see the Control Center interface, or select the Explorer watch face to see your signal strength at a glance.

>> **To check your cellular data usage:** Open the Apple Watch app on your iPhone, tap the My Watch tab, then select Cellular, and you can scroll to the Cellular Data Usage section.

>> **To enable or disable your cellular connection:** Simply swipe up from the watch face to open Control Center. Tap the green icon with the white wireless symbol inside, and then turn Cellular On or Off.

Got it? Great.

Monitoring the Apple Watch battery

One of the biggest challenges of such small technology that's always on? Battery life.

Thus, Apple gave its engineers a challenge to squeeze all-day performance out of Apple Watch. And they succeeded.

Okay, so *all-day performance* is a little vague, but Apple says it amounts to "up to 18 hours" — based on Apple's testing. This is even true for Apple Watch Series 5, and it's "always-on" Retina display.

Apple Watch includes a magnetic charging cable: One end snaps onto the back of the smartwatch and is secured by a magnetic connection — not unlike Apple's MagSafe charger for MacBook laptops — and the other end of the cable can plug into a computer's powered USB port (Mac or PC) or into a traditional electrical socket (with supplied adaptor on the end).

As you might've noticed — in this text or after looking in the box Apple Watch came in — the USB cable doesn't plug into Apple Watch anywhere. Unlike some other smartwatches, you have no port to uncover. Instead, the circular puck magnetically affixes to the underside of the watch, where the heart rate sensors are, and powers up the watch through induction technology.

The magnetic charger makes it easy to juice up Apple Watch because you don't have to open any ports on the watch to plug in a cable. Just attach it to the back of your Apple Watch, snap it into place, and you're good to go.

See Figure 2-7 for a look at the unique Apple Watch charger.

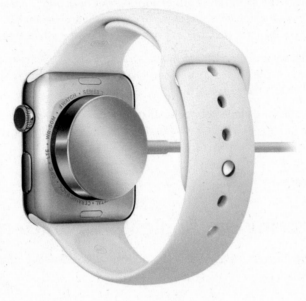

FIGURE 2-7:
Apple Watch
has no USB
ports — just
connect this
magnetic puck
to the back of
the watch and
plug the other
end into a
computer.

Broken down, the 18-hour battery life includes the following:

>> Ninety (90) time checks (four seconds long apiece)

>> Receiving 90 notifications

>> Forty-five (45) minutes of app use

>> A 60-minute workout with music playback from Apple Watch via Bluetooth

Newer and larger Apple Watch models typically experience longer battery life. The company also cautions that "battery life varies by use, configuration, and many other factors; actual results will vary."

If you're curious about specific tasks, Apple breaks down battery performance even further (based on its testing):

>> **Talk time test:** Up to three hours. Apple Watch was paired with an iPhone during the call.

>> **Audio playback test:** Up to 6.5 hours (when paired via Bluetooth with an iPhone).

- » **Workout test:** Up to 6.5 hours. Apple Watch was paired with an iPhone and had a workout session active and the heart rate sensor turned on.

- » **Watch test:** Up to 48 hours. This test was composed of five time checks per hour (each for four seconds).

- » **Power Reserve:** Up to 72 hours. Apple says its watch automatically switches to Power Reserve mode — when Apple Watch has only 10 percent battery remaining — so you can see the time up to 72 hours.

BATTERY-RELATED ACCESSORIES FOR APPLE WATCH

If you're on the go and don't want to worry about plugging the watch in somewhere to juice up, a number of power-centric products can help.

One called the Nomad Pod Pro (about $25) is a portable battery charger for the Apple Watch (and iPhone), featuring a 6,000-milliamp (mAh) battery. Nomad says this solution can provide multiple charges for Apple Watch — before the Pod Pro itself needs charging up.

Available in space-gray aluminum, the Pod Pro powers up the watch through magnetic induction, and a USB port is used to power up the Pod for when you need it.

(continued)

(continued)

Another option is the Kanex GoPower Watch Plus ($80), shown on the right. It's a powerful and portable 5,200 mAh battery with an integrated magnetic inductive charger for Apple Watch. Just like the Nomad product, this battery can charge Apple Watch and iPhone simultaneously. Kanex says you can charge an Apple Watch up to eight times before the GoPower Watch Plus needs recharging. The unit holds the magnetic watch charger in a cradle where you can feed the charging cable through to the back of the stand to reduce clutter.

As Apple suggests, you likely want to charge the watch at the end of each day to get it ready for the following one. If you only use the watch to tell time, you can get two or three days out of the watch without having to charge it (but you can't do anything other than see the time in Power Reserve mode). Should you want it, more information about battery life is available at apple.com/watch/battery.html.

Understanding the Home Screen

Although you can customize what app you see first, by default, the clock is what appears when you look at the watch. If you press the Digital Crown button, however, you can access your Home screen to see all the apps installed on the watch.

Apple Watch's Home screen, shown in Figure 2-8, is similar to other iOS devices —
namely, iPhone, iPad, and iPod touch. Basically, you'll see a bunch of icons that
launch an app when you tap one.

FIGURE 2-8:
While they're
bubbles
instead of
square icons
(on iPhone
and iPad), your
Home screen
apps should
still look
familiar to you.

The Home screen apps include Apple's own first-party applications, such as a
speech bubble for messaging or a cloud with a sun peeking behind it for weather.
In addition, you'll see any third-party apps you choose to install and transfer to
the watch from your iPhone; these might include a social media feed, such as
Twitter, news information provided by CNN, or a game, such as Trivia Crack. You
can see many of the native apps for Apple Watch in Table 1-1 in Chapter 1.

But instead of the neatly arranged rows and columns of app icons you see on an
iPhone or iPad, Apple Watch apps are arranged like bubbles in different sizes that
move as you press and slide your finger around the Home screen. The apps grow
larger when the icons get closer to the center of the Apple Watch screen, making
them easier to tap and launch.

If you twist the Digital Crown button while on the Home screen, you can zoom in
and out of all the apps you have installed on the watch.

TIP

If you zoom enough on one app, Apple Watch launches it for you! Also, while you're in the Clock app, you can simply press the Digital Crown button to return to the Home screen.

You can choose which apps are installed on Apple Watch via the Apple Watch app on your iPhone (preinstalled with iOS 8.2 and newer). An Apple Watch App Store is also built into the Apple Watch app for iPhone to allow you to download and manage your apps there (or you can download apps to the Apple Watch directly, if you prefer). See Chapter 11 for installing third-party apps.

See Figure 2-9 for a look at the Apple Watch app on iPhone.

FIGURE 2-9: Tap App Store inside the Apple Watch app on your iPhone to download new apps (left). Featured apps appear divided by theme, games, news, and other categories (right).

If you hold the Digital Crown button down for more than a second, it launches Siri, your personal voice-activated assistant (or simply say "Hey, Siri" into your watch). Only a quick tap on the Digital Crown button is necessary to open your Home screen. See Chapter 7 for more on using Siri on your Apple Watch.

Maintaining Your Apple Watch

Safeguarding your Apple Watch investment is probably one of the wisest actions you should take. After all, unless you received it as a gift (lucky you), you probably spent many hundreds on your Apple Watch, so it's a good idea to proactively protect it from damage.

Because your smartwatch is wearable, at least you don't have to worry so much about dropping it — like you would a smartphone, tablet, or laptop — but you still have steps you should take to ensure your Apple Watch runs smoothly for many years.

Avoiding water for the original Apple Watch

The first tip is to be mindful of water, but only if you own the original Apple Watch (Series 1), which debuted in 2015. This particular first-generation Apple Watch isn't waterproof. It might be splash-resistant, but you shouldn't fully submerge your wrist in water while wearing it. Keep your wrist away from the faucet when washing your hands or doing the dishes, and don't forget to take it off when you hop into the shower, climb into a bath, or step into any body of water. Your existing watch might be waterproof — and old habits die hard, as they say — so don't forget to take off Apple Watch before you do anything involving water. You really don't need to see who's texting you while you're soaking in a hot tub.

Getting caught in the rain or sweating profusely during a run are okay, says Apple, but err on the side of caution to be extra safe and leave Apple Watch behind before you go boating on a lake. Jogging with Apple Watch on a cloudy day? Don't worry about an umbrella because rain won't harm your gadget.

If you own Apple Watch Series 2 and newer, however, then it is indeed waterproof, and your device goes into Water Lock mode once it senses moisture! But here's something you should consider after a swim: Turn the Digital Crown to unlock the screen and you may hear sounds emanating from the Apple Watch, and you may feel some water on your wrist from the teeny speaker. To manually clear water from your Apple Watch, swipe up on the bottom of the watch face to open Control Center, then tap Water Lock. Twist the Digital Crown to unlock the screen.

Avoiding extreme temps

Be cautious when using your Apple Watch in extreme temperatures. This may be a relevant consideration based on where you live or work. Are you reading this book while lying by the pool in Acapulco or Maui? Or perhaps you work in the Athabascan oil sands up in northeastern Alberta, Canada?

Apple hasn't said what the optimum environmental requirements are for Apple Watch, but assuming for a moment they're similar to its iPhone, iPod touch, iPad, and MacBook laptops, the following should be kept in mind:

>> **Operating temperature:** 32° to 95° F (0° to 35° C)

>> **Non-operating temperature:** -4° to 113° F (-20° to 45° C)

>> **Relative humidity:** 5 percent to 95 percent noncondensing

To be sure, however, check Apple's website or ask someone who works at an Apple Store.

Doubting its durability

Although Apple advertises how strong its smartwatch is — between its solid materials, reinforced glass, and durable wristbands — try to remember you've got a sophisticated computer on your left or right wrist.

Remove the watch whenever you're doing something that could potentially damage it, such as tossing around a hardball with a child in a park or working under your car in a garage. Accidents happen, sure, but avoid the chance of impact to the area or harmful fluids from dripping onto the watch.

And then you should consider the bumps and knocks of everyday life: slamming a school locker shut, wrestling with your dog, or even putting away the dishes in drawers and cupboards. Don't be afraid to wear and use Apple Watch, but it's not immune to damage.

Considering a bumper

Just as other Apple products have countless accessories, you can bet Apple Watch is starting to see its fair share of optional add-ons to help you get more from the wearable — and this includes protective *bumpers*.

One of the most affordable is Modal's Apple Watch Bumper watch case — at only $10 apiece.

As you might suspect (or glean from the image in Figure 2-10), the rubber Bumper has been designed to snugly fit and protect your Apple Watch and make it action-proof — hence, the name of the company.

FIGURE 2-10: At $10 each, the affordable Apple Watch Bumper case from Modal is designed to protect your investment.

Although the Bumper wraps around the body of the case, you still have access to the Digital Crown button and the side button on the side of the watch. It also allows for full access to the screen, backside sensors on your skin, and unobstructed use of the microphone and speaker.

Using Apple Watch responsibly

Even though much of the advice I dispense is common sense, you'd be surprised just how often it's forgotten or ignored. Thus, the following are some suggestions to being smart about your smartwatch.

Watching the road, not your wrist

Just as you shouldn't be distracted with other technology while behind the wheel — holding up a smartphone, glancing at a tablet, or fiddling with a GPS navigation device — it's critical you resist accessing Apple Watch while driving a five-ton vehicle. As the late great Jim Morrison once famously sang: "Keep your eyes on the road — your hands upon the wheel."

"But I don't have to hold a smartwatch," you say.

True, but you can still tap the screen, press one of its two buttons, look down to read something, or hold the small speaker up to your ear to listen to Siri's voice — all of which could temporarily, even fatally, distract you when you should be concentrating at the task at hand.

Of course, we all know this, but look over to your left or right while at a stoplight and you'll no doubt see someone using his or her tech gadget while in the driver seat.

It may be temping, but it can wait. The more technology we have at our fingertips, the more likely we may want to use it wherever and whenever.

Even using hands-free technology has been proven to distract drivers, so although you may be looking at the road and keeping your hands at "10 and 2" (as your driving instructor likely taught you), not exclusively focusing on your driving could be an issue.

Not here to lecture, of course, but even if this section reminds you about the often-overlooked dangers of distracted driving, it did its job!

Watching out for sidewalks too

Anyone who's spent time on YouTube might've seen some humorous videos of people walking down the sidewalk, oblivious to the world around them because they're staring at their smartphone, and as a result, they walk into walls or people, trip on a curb, or even fall into a manhole.

Funny? Sure. These clips prove not all of us can multitask as well as we think, especially because we can't be looking at our phone and what's in front of us at the same time.

But what you might not see on YouTube are distracted pedestrians walking into or across a road and getting hit by a vehicle. That's not so humorous, although you may be tempted to nominate these unfortunate souls for a Darwin Award, a silly online commemoration of people who perish in ridiculous ways to protect our gene pool — a nod to Charles Darwin's evolutionary theory.

One tiny little mistake in judgment and you could be seriously injured or killed — or force a car to swerve out of the way and injure passengers or other pedestrians. It's happened.

Therefore, while Apple Watch was meant to be worn and used, be mindful about your surroundings — even when you're on foot. If you want to look at, talk to, or hear your wrist-mounted companion, stop walking first.

Remembering netiquette — even when not on the Internet

Ever been to a restaurant and looked around at the other tables? You might have noticed something peculiar over the past few years: People are looking more at their mobile devices than the people they're with.

Perhaps it's an unfortunate sign of the times — be it a date night disturbed by someone checking the score of his or her favorite team, or a group of friends who'd rather advertise where they are to people they're not with than appreciate those they're sitting beside.

Will the same thing happen with smartwatches? Or maybe wearable technology will be more discreet than a smartphone because you can casually glance down at your wrist while sipping a drink instead of navigating through menus on a 5.5-inch handheld?

For the sake of humanity in the digital age, we should hope technology helps rather than hinders human interaction — whether it's a couple enjoying a quiet dinner in a restaurant, kids sitting in a classroom, or business associates collaborating on a project.

Technology has a time and a place — and smartwatches are no different.

Reducing the likelihood of theft

Another take on discretion: Because many can't afford an Apple Watch, you might think twice about flaunting it on your wrist. Just as millions of smartphones are stolen in the United States each year, 33 percent of them are lost or broken (says DMR, or Digital Marketing Ramblings).

Many articles have been written about the dangers of tweens and teenagers showing off their newly purchased gadgets, such as smartphones and expensive headphones, so try to be a little more discreet with your Apple Watch. The last thing you want to do is tempt fate and lose your smartwatch forever — or, worse, put your life at risk over a mere gadget.

Also, while Apple Watch doesn't hold a lot of data on it — most of it is stored on your iPhone — you still don't want your personal property falling into the wrong hands.

This might all seem a little too preachy, but I simply want to get some of these obvious — or perhaps not-too-obvious — safety, privacy, and common sense tips out of my system because I see the unspoken rules broken all the time.

Taking Advantage of Accessibility Features

In Chapter 11, I cover how to install third-party apps and tweak Apple Watch settings from inside the Apple Watch app on your iPhone. I also look at the various accessibility options and how to enable them.

But consider this section a short preview of what you can do with Apple Watch's accessibility options, which refer to features designed for those with visual or aural impairments.

Not unlike myriad options on iOS (iPhone, iPad, and iPod touch) and OS X for Mac, Apple Watch comes loaded with accessibility options you can enable in the Apple Watch app on the iPhone.

Specifically, you can enable these options:

>> **VoiceOver:** A screen reader available in one of 14 languages.

>> **Font Adjustment:** Adjust (enlarge) fonts in many apps, such as Mail, Messages, and Settings.

>> **Bold Text:** Enhancing text to make it easier to read.

>> **Extra Large (X-Large) Watch Face:** A watch face with bigger hour/minute numbers.

>> **Zoom:** Easily magnify information on the screen.

>> **Grayscale:** Strips color from various apps and menu screens.

>> **Reduce Transparency:** Reduces the contrast to make text easier to read.

>> **On/Off Labels:** Makes it easier to access settings to enable or disable features.

>> **Reduce Motion:** Home screen is easier to navigate.

>> **Mono Audio:** Sends same-channel audio to both ears, or you can increase/decrease the volume in one ear.

>> **Taptic Engine:** A "linear actuator" inside Apple Watch produces haptic feedback, which is a slight vibration on your wrist every time a notification comes in.

>> **Chimes:** If enabled, you can hear a chime at a specified interval (such as hourly) and can select the kind of chime you want to hear.

>> **Scribble:** If you're not using the Voice Dictation feature and you can't find the right preset "Smart Reply" option to a message, you can write back by scribbling letters on the display. Apple Watch will convert your scribble into text for you.

>> **Motor:** Select the side button click speed (default, slow, slowest).

>> **Touch Accommodations:** If you have trouble using the touchscreen, adjust some settings here to change how the screen will respond to touches.

>> **Activity (Wheelchair):** Since the first Apple Watch (Series 1), Apple has optimized the Activity app for wheelchair users, taking into account different pushing techniques (for varying speeds and terrain). In addition, you can find two wheelchair-specific Workout options and a Time to Roll notification option. Figure 2-11 showcases these options.

FIGURE 2-11: Apple Watch's many accessibility features include an Activity app optimized for wheelchair users (left), which can remind you to move too (right).

IN THIS CHAPTER

» Recognizing the difference between a tap and a press

» Swiping the Apple Watch screen

» Exploring the Digital Crown button and the scroll feature

» Using the side button

» Going hands-free with Siri

» Understanding tactile feedback: Apple Watch can tap *you*

» Previewing the built-in Apple Watch apps

Chapter **3**

Control Freak: Mastering Apple Watch's Interface and Apps

Are you ready for a deeper dive into Apple Watch's controls, features, and options? That's precisely what I cover in the following pages — or digital pages if you're reading *Apple Watch For Dummies* in ebook form.

Though it's unlikely you'd want to read an electronic book on Apple Watch's diminutive screen, your always-on mobile companion can do so much to help you throughout the day. This chapter explores what's possible, beginning with mastering the controls.

As with many other Apple products, the user experience is paramount, and the Cupertino, California–based company has nailed the interface once again with Apple Watch.

In other words, Apple Watch is easy to use.

After you look at using Apple Watch's screen and buttons (and microphone and speaker), I suggest a few different ways to take advantage of the watch's main features — including Dock, Control Center, and Notifications — and then I wrap up with an overview of the built-in Apple apps.

TIP

Although some of the Apple Watch models are best used while wirelessly tethered to a nearby iPhone, the watch does have 16 gigabytes (GB) of storage you can use (or half that for older models). Well, not all of it, mind you, because Apple uses the integrated storage for various things, including watchOS (operating system) and parts of installed apps. But you can store quite a bit of offline music and podcasts; therefore, you can leave your phone at home if you go for a jog (although Bluetooth headphones, such as Apple's trendy AirPods, are recommended). See Chapter 9 for more on music playback. You can also upload a lot of photos for viewing if your phone isn't around.

Handling Apple Watch's Controls

This watch is on your wrist, so the main way you interface with it is with your fingertips. But you have a few different ways to do it.

Just like with your iPhone, iPod touch, and iPad, you can use your fingers on the Apple Watch screen to tap, double-tap, press, two-finger press, and swipe. The Digital Crown button and the side button also help you access myriad features on your Apple Watch.

Tap

Tapping on something, such as an icon, is akin to pressing Enter on a computer, or clicking the left button on a computer mouse. Tapping confirms a command, as in "Yes, I want that" — whether it's playing a song, accepting a calendar invitation, starting turn-based directions on a map, or hanging up on a completed call. As shown in Figure 3-1, tapping is the frequent method for interacting with Apple Watch.

Speaking of taps, you can also send a *tap* to friends and loved ones to let them know you're thinking about them. Apple calls this its *Digital Touch* feature — a communications option that lets you send an image and subtle vibration to someone else's Apple Watch (required). See Chapter 5 for more on sending taps to friends and loved ones.

Double-tap

You might not use double-tap very much, but Apple Watch does recognize this action. First, you must enable Accessibility settings by triple-tapping the Digital Crown button (see Chapter 2 for more on these settings). Once you do so, you can activate *Voice Over*, which is having text spoken to you in a humanlike voice, simply by double-tapping the screen. Another Accessibility option is Zoom: Quickly tap twice on the screen with two fingers for a closer look at text or images.

Press (Force Touch)

Apple Watch understands the difference between a light touch and a deep press and reacts accordingly. That is, you can quickly tap, as explained in the "Tap" section, or press with your finger and hold for a second to trigger access to a range of contextually specific controls tied to the task at hand. Apple calls this *Force Touch*. When you use Force Touch, pressing firmly on the screen displays additional controls in such apps as Messages, Music, and Calendar. It also lets you select different watch faces, pause or end a workout, search an address in Maps, and more. Force Touch is the most significant new sensing capability since Multi-Touch on smartphones!

Two-finger press (time or heartbeat)

If you lightly press two fingers on the Apple Watch clock face screen, you hear a voice announce the time for you!

Also, you can let someone special know you're thinking about him or her by sending that person your heartbeat, as shown in Figure 3-2. The catch? He or she also needs an Apple Watch.

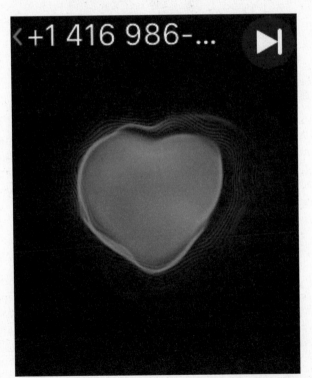

FIGURE 3-2:
Send your heartbeat by pressing two fingers on the screen at the same time.

In the Messaging app (Chapter 5), tap the Heart icon and then press two fingers on the screen at the same time and the built-in heart rate sensor records and sends your heartbeat to a loved one, who will feel it on his or her wrist. Apple calls it "a simple and intimate way to tell someone how you feel," and it's a feature not found on any other smartwatch.

Again, please flip to Chapter 5 for more on sending your heartbeat to someone!

Swipe

Like most other mobile screens in your life — such as a smartphone, tablet, and many laptops — Apple Watch also supports onscreen swiping. If the app supports it, you can swipe around using your fingertip.

Sketch

Just as you can tap, press, and swipe on the Apple Watch screen, you can also use your finger to quickly draw something on the small display and then send it to someone else who wears an Apple Watch. Apple calls this a *sketch*. It could be a flower, heart, puppy, butterfly, wedding ring, Christmas tree, or whatever. See Figure 3-3 for an example of a sketch.

FIGURE 3-3: Sketch whatever you like — in your desired color — and then send it to another person's Apple Watch.

Not only will the person you send your sketch to see what you draw on his or her Apple Watch, but that person can also watch your sketch come to life with animation — and then respond back with a sketch of his or her own. See Chapter 5 for more on sketches.

Digital Crown button

Apple is known for introducing new ways to interact with content, and the Digital Crown button is no exception. If the watch is worn on the left wrist, this button is on the top right side of the Apple Watch case. It looks like the crown on mechanical watches, which was traditionally used to wind the main spring and to set the time. For the watchOS platform, it's used primarily to magnify content on the small screen without your fingers getting in the way of content. Instead of pinching to zoom, as you would on an iPhone or iPad, you twist the Digital Crown forward or backward to zoom in and out of photos or maps. You can also use it to quickly scroll through contacts, songs, and more.

Keep in mind that you can flip the band around to wear the watch on your right wrist, which places the Digital Crown on the left-hand side of Apple Watch. Of course, you can keep the side button and the Digital Crown button on the right side of the watch while wearing the watch on your right wrist, but it may not be comfortable for you to access these buttons with your left hand. But be sure to go to the Settings area to change which wrist you're wearing your watch on. See Chapter 11 for more on changing the watch's orientation.

At any time, you can also press the Digital Crown button to see the watch face or (if on the watch face already) return to the Home screen (which is similar to pressing the circular Home button on an iPhone or iPad). Or press and hold the button to activate Siri. Double-pressing the Digital Crown button switches between the watch face (the one you've chosen to display) and the last app you used. It's a fast and convenient way to see the time, regardless of the app you're in. Twist the Digital Crown to zoom, scroll, or adjust what's on the screen.

To summarize, use this list to determine how to use the Digital Crown button for a specific task:

>> **To return to the Home screen:** Press the Digital Crown button once to return to the watch's Home screen (similar to pressing the circular Home button on an iPhone or iPad).

>> **To return to the clock/to return to the most recent app:** Double-press the Digital Crown button to see the clock again. Double-press the Digital Crown button again to go back to the last app you were in. This is a fast and convenient way to see the time regardless of what app you're in.

>> **To activate Siri:** Press and hold the Digital Crown button to launch your voice-activated personal assistant. Or you can just say "Hey, Siri" into your watch, followed by a question or command.

>> **To zoom/scroll:** On the Home screen and in supported apps, twist the Digital Crown button forward or backward — as if you were winding your watch — to scroll through lists or zoom in on a photo, a map, the Home screen, and more.

>> **For swimmers:** On Apple Watch Series 2 or later, turn to unlock the screen after a swimming workout.

Side/Power button

Apple Watch has another handy button that's located just below the Digital Crown button on the right side (if you're wearing Apple Watch on your left wrist). Depending on whom you ask, it's called the side button or the Power button.

REMEMBER

You can wear Apple Watch on the right wrist and turn around the watch case, which puts the side button (and Digital Crown) on the left side of the watch. See Chapter 11 for more on changing the watch's orientation.

Some refer to this side button as a Power button because holding it down for a couple seconds displays a power-down screen where you can choose to turn your watch off. This is similar to pressing and holding the Power button on the top or side of an iPhone or iPad.

Double-pressing the side button initiates Apple Pay, thus allowing you to use your Apple Watch to make a purchase at a participating retailer or compatible vending machine. Once you pair your Apple Pay account with a credit or debit card, you simply wave your NFC-enabled (near field communication) smartwatch over a contactless sensor to initiate the transaction. You should hear a faint tone and feel a slight pulse vibration to confirm this *digital handshake* has been completed. Apple Pay works on Apple Watch as long the watch stays in contact with your skin. See Chapter 10 for more on Apple Pay.

Finally, pressing and holding the button initiates the SOS emergency feature, which allows you to quickly and easily call for help and alert your emergency contacts that you're in danger.

The following is a summary of the side button's features:

>> **Power:** Press and hold the side button until you're prompted to turn off the power or activate SOS. You probably won't use this very much — just like you don't power down your iPhone or iPad often. But you can if you want.

>> **Pay:** Double-tap the side button to launch Apple Pay for when you're about to buy something. Remember, Apple Pay requires skin contact to operate, so ensure the watch is snug on your wrist before you press the side button twice to launch Apple Pay.

>> **SOS:** Press and hold the side button to display an option to power off the watch or activate SOS, if you need to have your watch call an emergency contact or 9-1-1. If you don't have a cellular Apple Watch option, you require a nearby iPhone for these emergency numbers to be dialed. Chapter 8 has more on the SOS feature.

Going Hands-Free with Siri

Can you think of an even more natural way to interface with a smartwatch than touching it?

How about talking into it?

Apple's Siri (pronounced "sear-eee") is a voice-activated personal assistant that lets you ask questions or give a command on an iPhone, iPad, or iPod touch — and now Apple Watch. As with the other iOS devices, you simply press and hold the circular Home button to ask a question.

In Chapter 7, I cover all the different ways Siri can help you master Apple Watch, but for now, I offer a quick snapshot on how to use it.

To activate Siri on your Apple Watch, follow these steps:

1. **Press and hold the Digital Crown button.**

    Siri is ready for your instructions after the short chime.

2. **Ask Siri a question or give a command.**

   You could, for example, ask "Who's winning the New York Yankees game?" In fact, you don't have to say the full team's name; therefore, asking how the "Yankees" are doing rather than the "New York Yankees" is usually fine. Siri then shows you the requested information or tells you what you asked for.

   As the information is being displayed — in real time, no less — Siri also says something like "Okay, sports fans, let's have a look," or "Here you go."

3. **Press the Digital Crown button again if you want to go back to the Home screen.**

   You should then see the Apple Watch's main screen — flush with icons.

REMEMBER

Just like Siri on an iPhone, iPad, or iPod touch, you need an Internet connection to use Siri on Apple Watch. The requested information is sent to Apple's servers to process — like your question about the baseball score — and is then sent to your watch to give you the answer. Unless you have the cellular model and have set up your monthly plan, your watch needs to be connected via Bluetooth or Wi-Fi to a nearby iPhone.

Vibrating Along with Apple Watch's Tactile Feedback

Apple Watch can be tapped, pressed, swiped, and spoken to. But your smartwatch can also tap *you* in the form of a light vibration. Powering this technology is what Apple calls a *Taptic Engine*.

Apple's Taptic Engine is a *linear actuator* inside Apple Watch that produces haptic feedback; in plain English, it's a slight *buzz* on your wrist whenever you receive an alert or notification or press down on the display (like some smartphones offer). Another comparison might be a video game controller that buzzes in your hands in conjunction with what's happening on your TV screen (like your character being shot in a first-person shooter).

Not only does Apple Watch's Taptic Engine give you information without your having to even look down at your wrist — such as Apple Watch telling you (based on the number of taps you felt) to turn left or right while using Apple Maps — but it also enables new and intimate ways to interact with those who also own an Apple Watch. Chapter 5 further discusses how to send a tap or your unique heartbeat to a friend or loved one.

Using Control Center, Dock, and Notifications

Apple Watch gives you a couple ways to glean information while on the go: Control Center, Dock, and notifications.

Remember, Apple Watch isn't meant for reading lengthy websites — actually, the watch doesn't have a web browser — because it's designed for quick interactions. Though they differ, Control Center, Dock, and notifications give you bits of customized information when and where you need them.

Control Center

On an earlier version of watchOS, swiping up from the bottom of the screen would launch Glances, which were quick snippets of information, such as how your favorite sports team was doing, weather information, or how your stock was performing. Now, this motion initiates the Control Center — a quick look at information on your watch, for example, such as battery percentage, the Wi-Fi and cellular signal, flashlight mode (turning the screen bright to see in the dark), Airplane mode, and more than a half-dozen other options (see Figure 3-4).

FIGURE 3-4: Access the Control Center on Apple Watch (left) by swiping up from the bottom of the screen. Keep swiping to see different icons that can be used to quickly make changes (right).

To activate Control Center on your Apple Watch, follow these steps:

1. **Swipe up from the bottom of the clock (time) screen.**

    You'll see common icons at a glance. Common icons include:

    - Remaining battery percentage

    - If your Apple Watch is using Wi-Fi or cellular

    - Whether your flashlight app is on (helps you see in the dark, with three options)

    - If Water Lock is enabled, which then pushes water out of the speaker after swimming

    - Pinging iPhone (to make your nearby phone ping/ring loudly so you can find it)

    - Whether Airplane mode is on

    - Whether your Theatre mode is activated (this mode keeps the watch silent and the screen dark until you tap it)

    - Choose where to play audio (Apple Watch or a wirelessly connected device, such as AirPods)

    To see what this looks like on Apple Watch, see Figure 3-5.

2. **Select one of the Control Center options.**

    Tap one of the dozen or so options and tap again to turn it off. Can't find your iPhone? Tap the icon that looks like a phone with sound coming out of it and your Apple Watch will audibly "ping" your missing iPhone to help you locate it nearby (with Bluetooth it's generally 30 to 50 feet).

3. **Tap Edit to customize your Control Center.**

    You can change what you see on your Dock by scrolling to the bottom of the Control Center and selecting Edit. Now, the icons will wiggle, which means you can move them "drag and drop" style into a desired order on the Apple Watch screen.

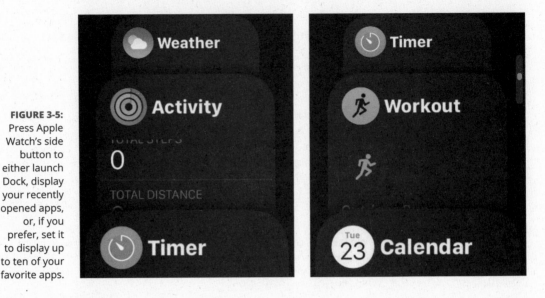

Dock

Over the last couple of years, Apple added a handy Dock feature for Apple Watch wearers. You activate this by pressing the side button. Quite simply, Dock lets you quickly open your favorite apps or move from one app to another.

Here's how to get going:

1. Choose which apps you want to appear in Dock.

You can select up to ten of your favorites, in fact. To choose what you want, simply:

a. *Open the Apple Watch app on iPhone.*

b. *Tap My Watch and then choose Dock.*

c. *Tap Edit and then add or remove apps to choose your favorites.*

To rearrange apps, touch and hold next to an app, then drag up or down.

d. *Save your changes by tapping Done.*

2. Press the side button to activate Dock.

3. Swipe up or down.

Alternatively, you can turn the Digital Crown. This will cycle through the last apps you opened or your favorite apps.

4. **Tap to open an app.**

If you scroll all the way down to the bottom of the screen, you can tap All Apps to go to the Home screen.

5. **Close Dock by pressing the side button again.**

Notifications

Much like notifications on an iPhone or iPad, Apple Watch notifications display app information at a specific time. This could be a voicemail waiting for you, a calendar appointment, a news headline from CNN, a social media notification (someone started following you or liked your photo), a comment made to your online classifieds ad, or a game letting you know it wants you to come back.

Apple Watch has two kinds of notifications:

>> **Short-look notifications:** These appear when a local or remote alert first arrives (see Figure 3-6). They present a minimal amount of information to you. If your wrist is lowered, the short-look note disappears. Information includes the app name, icon, and title.

>> **Long-look notifications:** These appear when your wrist remains raised or if you tap the short-look interface. You receive more detailed information and more functionality, such as four action buttons for additional information. You need to dismiss these notifications when you're done by swiping up.

Apple asks its app developers to make sure notifications are relevant to what the user wants and to not bombard them with messages all day long. Nonetheless, you can turn off notifications for any app in the Apple Watch app on iPhone; see Chapter 6 for more on how to do this.

FIGURE 3-6:
Notifications on your Apple Watch can show you a bit of time- or location-relevant information.

Looking at Apple Watch's Built-In Apps

Similar to the preinstalled apps designed to help get you started on other Apple products, Apple Watch has a number of built-in apps too. These include:

>> **To keep you in touch:** Phone, Messages, and Email

>> **To prevent you from getting lost and to find stuff around you:** Maps

>> **To stay organized and informed:** Calendar, Clock, Weather, and Stocks

>> **To entertain you:** Music and Photos

I briefly listed some of these apps in Table 1-1 in Chapter 1, but here, I discuss all of them in greater depth because you're likely to rely on many of these apps. Okay, so some might not appeal to everyone, such as a Stopwatch or a World Clock, but they're available if, and when, you want them.

Phone

You no longer need to reach for your iPhone to see who's calling you. Simply glance at your wrist to see the name or number (if he or she isn't in your Contacts) and decide whether to take the call. You can have a chat through the watch if you like, or transfer the call over to your iPhone. Don't want to take the call? Simply cover Apple Watch with your hand to mute it. You can also place a call to someone through Apple Watch — even if your iPhone isn't nearby — providing you have a cellular plan for your supported Apple Watch (or you make calls over Wi-Fi). See Chapter 5 for more on talking on your Apple Watch.

Messages

If someone sends you a text message (SMS) or an iMessage to your iPhone, Apple Watch gives you a subtle tap to let you know about it. Raise your wrist to see who wrote it and to read the message. You can reply with a preset response, send an animated emoji, dictate the response, or record and send a short audio message. See Chapter 5 for more on handling messages on your Apple Watch. Figure 3-7 shows an example of a text message.

FIGURE 3-7:
Because
the Apple
Watch has no
keyboard, you
have to use
your voice to
dictate a reply.

Mail

Although a small smartwatch screen may be more conducive for short text messages than lengthy emails, you can read your personal or professional email on
your watch, as shown in Figure 3-8. You can also flag emails, mark them as read
or unread, reply to them, or delete them (press and hold on the screen to access a
few options. But you can't create an email from scratch (you'll need your iPhone
for that). See Chapter 5 for more on reading and responding to emails on your
Apple Watch.

FIGURE 3-8:
Read your
email on Apple
Watch.

Calendar

Your wrist can tell you when you've got an upcoming calendar appointment. You can set meeting reminders, accept or decline calendar invitations, and, if desired, email the organizer with a preset response. See Chapter 6 for more on Calendar options.

Alarm

Apple Watch lets you set and manage multiple alarms, such as those for meetings or wakeup calls. You can either ask Siri to set them or set them yourself using the Digital Crown button to tweak the alarm time. You can even choose for the alarm to vibrate your wrist. Figure 3-9 shows the Apple Watch's Alarm app. Don't forget you can also sync your iPhone alarms to your Apple Watch. See Chapter 4 for more on setting alarms.

FIGURE 3-9:
Use the Alarm app to have your watch wake you up.

Stopwatch

Much like the stopwatch on your iPhone, Apple Watch offers one for your convenience, and you can set it to digital, analog, or hybrid view. Apple Watch also offers an optional graph view. See Chapter 4 for more on using the Stopwatch app on your Apple Watch.

Timer

Whether you're running around a track or cooking something in the oven, Apple Watch has a timer for that. As the timer runs, you should see a line move around the dial to give you a sense of how much time has passed and is left. See Chapter 4 for more on setting timers on your Apple Watch.

World Clock

You don't need to go to a website to see the time in different cities around the world (or count on your fingers as you do the manual calculation between, say, New York and London). Launch the World Clock app to see the time in cities of your choosing. You can use your iPhone to add new locations at any time. See Chapter 4 for more on setting up the World Clock app.

Siri

Siri is a fast and convenient way to interface with your Apple Watch. In fact, nothing is more natural than using your voice; your smartwatch will respond after you say "Hey, Siri," followed by a question or command. Alternatively, you can push and hold the Digital Crown button to activate Siri. See Chapter 7 for more on using Siri to help you do things on your Apple Watch.

Weather

Use Apple Watch to check the weather where you are or for any city in the world. This app shows the temperature and precipitation at that exact moment or for the days or week ahead. See Figure 3-10 for a glimpse at the Weather app. If you have not chosen a watch face that includes weather information, the quickest way to access that information is to ask Siri for it. The second-quickest way is to swipe up and access the Weather Glance screen. Of course, opening the app itself is the third-quickest way, but you should choose a method that works best for you. See Chapter 6 for more on the Weather app on your Apple Watch.

FIGURE 3-10:
We love knowing the weather, and Apple Watch provides multiple ways to see it, including via the Weather app.

Stocks

Follow all the companies you have a vested interest in — or plan on buying into — with the Stocks app on your Apple Watch. This app lets you keep up with the stock market price, point and percentage, market cap, and more. All your stocks include a historical graph. See Chapter 6 for more on the Stocks app on your Apple Watch.

Activity

Activity is one of the two important fitness-related Apple Watch apps. The three Activity rings — Move, Exercise, and Stand — give you a simple yet informative glimpse into your daily activity goals and progress. The app was designed to motivate you to sit less and move more. See Figure 3-11 for the three Activity rings, and see Chapter 8 for more on the Activity app on your Apple Watch.

FIGURE 3-11:
This smart, color-coded Activity app shows your daily progress.

Workout

During an exercise routine, the Workout app shows you real-time fitness information, including time, distance, calories, pace, and speed. You can choose from preset workouts, such as Running, Walking, Cycling, and others. Figure 3-12 shows you the Workout app. Chapter 8 gives more information on how to use it on your Apple Watch.

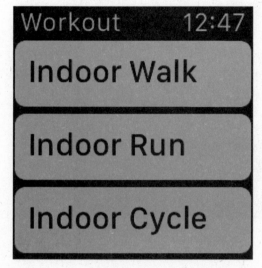

FIGURE 3-12:
The Workout app offers some exercise routines.

Maps

Whether you're trying to find a restaurant in your hometown or you feel like going on a stroll in an exotic city, your Apple Watch can give you directions based on your current location, as shown in Figure 3-13. See the fastest route, get turn-by-turn navigation instructions (including taps on your wrist when it's time to turn), or ask Siri to find local businesses. See Chapter 6 for more on the Maps app on your Apple Watch.

Photos

Those who matter to you are now just a glance away. The Photos app on Apple Watch displays your photos of loved ones, friends, pets, scenery, and other memories. Use the Digital Crown button to zoom in on individual images or swipe to browse through them one photo at a time. You can even choose photos to load on your watch — even if your iPhone isn't near. See Chapter 12 for more on using the Photos app on your Apple Watch.

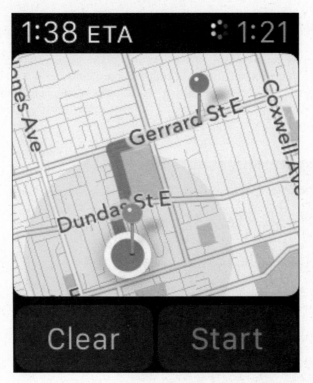

FIGURE 3-13:
Get directions
to a location,
such as a local
business,
by using the
Maps app.

Camera

Although Apple Watch doesn't have a camera, you can use it as a live viewfinder for your iPhone's rear-facing iSight camera. You can, therefore, use your watch to see and snap a subject, essentially turning your wrist into a wireless shutter (ideal for selfies). Plus you can adjust the timer remotely. See Chapter 12 for more on using your Apple Watch to control your iPhone's camera.

Music

You can keep your iPhone tucked away in your pocket or purse and play your music remotely on your Apple Watch. Figure 3-14 shows what your songs look like on your watch, complete with album artwork if you have it on your iPhone. You might want to load up your watch with a few hundred tunes to listen to when you don't have your phone with you (Bluetooth headphones are required). See Chapter 9 for more on listening to music or managing your music on your Apple Watch.

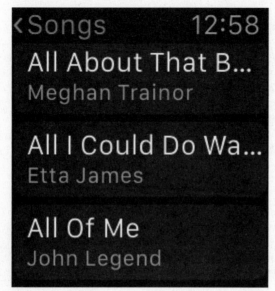

FIGURE 3-14:
Like music?
You'll love
Apple Watch
because it can
store music
you synced
from your
iPhone. You
can also
control tunes
on your iPhone
or stream
directly from
Apple Music.

Remote

Not only can you remotely control your iPhone music via your Apple Watch, you can also use it to access a nearby Apple TV box connected to your TV. Your wrist-based remote can navigate the main menu, scroll through media lists, and select what you want. Your watch's Remote app can also be used to control your iTunes library and iTunes Radio on a PC or Mac. See Chapter 9 for more on using your Apple Watch as a remote control.

Wallet and Apple Pay

Formerly referred to as Passbook, this handy app for iPhone also works well on Apple Watch, and lets you keep track of such things as boarding passes, movie or theater tickets, loyalty cards, and more. Time-based alerts let you know when you should or can use these plane or train tickets, loyalty cards, and such. When you shop with Apple Pay, you choose on which credit or debit cards to make the purchase. See Chapter 10 for more on using Apple Pay with your Apple Watch, as well as for more about Wallet.

Apple News

As you might expect, this convenient app curates some of the top news headlines for you to read on your watch. You might find yourself in line at a supermarket or waiting for a friend on a park bench; tapping this red icon lets you view some of the top headlines as you wait. If there's something that interests you but you don't have time to read it now, tap Save for Later, and the entire article will appear on the Apple News app on iPhone. Want to move on? Tap Next Story. News sources include Washington Post, Fox, CNN, Newsweek, Billboard, VentureBeat, and many others.

Breathe

Live a better day by taking a minute to Breathe. Figure 3-15 is what you'll see when you tap the teal-colored Breathe app on Apple Watch. "Quiet your mind, relax your body and be in the moment," it continues, before asking you to try and complete at least one session per day. You'll also be nudged to Breathe on your Apple Watch, when it thinks you should take a break, and you can twist the Digital Crown to change the duration of the breathing exercises the app guides you through.

FIGURE 3-15:
Take a moment out of your hectic life and focus on your breathing with the Breathe app.

Walkie-Talkie

Communicate Apple Watch to Apple Watch. This app lets you have a voice conversation with someone else via your watches. Any model of Apple Watch works — even Series 1 — but you need Apple watchOS 5 or newer to use it. Contacts you add, such as your partner, kids, or boss, can talk to you whenever you're set to "Available." You'll see some Suggested contacts (like family members) and then a list of all the people listed on your iPhone's Contacts app. Tap to invite them and they'll have a chance to accept. Swipe down to select if you're available to chat or not. There's a deeper dive of Walkie-Talkie in Chapter 5.

Heart Rate

As you might expect, tapping this little red heart shows your current heart rate and daily summaries on Apple Watch. This Health app can notify you if Apple Watch detects your heart rate is high or low. You can adjust or turn off these notifications in the Apple Watch app on your iPhone. Tap Turn On to activate it, and you'll see it measuring your beats per minute (BPM) and logging the time your rate was measured as well as and some historical info.

Find Friends

The little orange icon with two white stick people launches the Find Friends app. The first time you tap this icon, the app asks you permission to access your location on a map while it's in use and to allow people that you select to see you. The app then shows your location on a map and you (and they) can see how far away friends are. Tap Allow or Don't Allow when prompted (first time only).

Podcasts

In case you haven't heard, podcasts are free downloadable programs. These can be a comedy routine, news report, political rant, tech advice, religious sermon, or the latest music remix from a popular club DJ. Podcasts can be video, too, but only audio versions work on Apple Watch. When you subscribe to podcast, the show automatically downloads to your device so you can hear it on-demand. The Podcasts app on Apple Watch syncs with your iTunes library so you can hear everything from your wrist (presumably to wireless headphones). See Figure 3-16.

FIGURE 3-16:
Sync or stream your favorite podcasts right to your wrists — and then to your Bluetooth headphones!

Radio

Speaking of audio, Apple Watch's Radio app lets you stream music to your watch, either via a nearby iPhone over Wi-Fi, or through a cellular connection on some supported models; and you can customize the stations, of course. By default, the app lists some featured stations, but you can choose a desired genre (like Jazz or Hip-Hop), theme or region (like Workout, Indie, Kids, or from a specific country), or hear news — all delivered via audio.

Audiobooks

New with WatchOS 6, the Audiobooks app, shown in Figure 3-17, on Apple Watch will list and play all books on your iPhone (inside the Apple Books app). Neat, huh? That is, all titles in your Reading Now list are automatically synchronized to your watch. So now you can listen to a chapter while out for a jog and then continue on when you're back on your iPhone or iPad later, as it'll remember where you left off. After you open the Audiobooks app, you have options to how to play acquired audiobooks.

FIGURE 3-17: Play all your audiobooks from your Apple Watch.

App Store

You no longer need an iPhone to browse and download Apple Watch apps to your wrist! Simply tap the large A icon on your Apple Watch home screen to launch the Apple Watch App Store, shown in Figure 3-18, to start browsing (Trending, Featured apps) or searching (using Scribble or Dication), along with looking at screenshots, reading reviews, price, and more. It's about "time" we could access the App Store on Apple Watch directly!

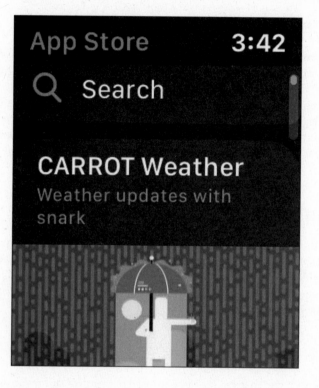

FIGURE 3-18: Browse or search and download new Apple Watch apps directly to your device.

Noise

High noise could damage your hearing. The new Noise app on Apple Watch has an ear out for you, as it can sense when ambient noise levels reach a dangerous level. Whether it's at a rock concert or working with heavy machinery, the app might suggest you put in some earbuds; tapping your wrist shows you the estimated decibel level and duration of your exposure. Chapter 8 has more information on the Noise app.

Cycle Tracking

Cycle Tracking is a new Apple Watch app (which I cover more in Chapter 8) that provides insight into a woman's menstrual cycle, to help paint a clearer picture of your overall health. Along with data you may want to share with a physician — flow information (Light, Medium, Heavy), symptoms (like cramps or headaches), and cycle length — you can see a graphical chart on your iPhone's Health app. The app can even alert you when it predicts that your next period or fertile window is about to start.

Home

Smart homes all around the world are increasingly powered by Apple HomeKit-enabled products, which means these supported devices all talk to one another securely; these products let you use an app or voice, through Siri, to control it all. For example, you can say "Hey Siri, turn on the lights" or "Lower the thermostat by two degrees." Apple Watch lets you control your smart home on your wrist too. Tap the orange and white Home button and all your HomeKit items are listed. Just remember that you first need to add the Home app on an iPhone to get things to work.

Calculator

Now with a dedicated app on your wrist, access a calculator to quickly calculate a tip, add some receipts or checks you're depositing into a bank machine, and more. The calculator has the familiar orange, black, and white virtual buttons as the iPhone version. (Don't forget you can also raise your wrist to activate Siri to ask a math question, as well!)

Voice Memos

Ever come up with a great idea you wanted to capture? Or maybe you need to record a meeting for work or a lecture for school? Whatever your reasons for wanting to record audio, you can now do so on your Apple Watch, thanks to the dedicated Voice Memos app. Simply tap to launch, press the large red Record button and tap again to stop. All recorded memos will be synced to your iPhone, iPad, and Mac.

Settings

The Settings app for Apple Watch lets you enable or disable a number of settings, including Airplane mode (turning off all wireless radios), Bluetooth, and Do Not Disturb. This app also mutes your watch in case you don't want to hear any sounds emit from it. Did you lose your iPhone under the cushions? Your Apple Watch can make your iPhone *ping* loudly so you can hear it and find it. See Chapter 11 for more on the Settings app on your Apple Watch.

Compass

New to Apple Watch Series 5 is a built-in "magnetometer" inside the watch that detects magnetic north, and then automatically adjusts to show true north, so you now know which way you're facing in the Maps app (this can be quite handy). The standalone Compass app also shows additional information — heading, incline, latitude, longitude, and current elevation — or you can have compass information as a watch face complication (see Chapter 4). Be aware, Apple says bands with magnetic clasps (opposed to leather or silicone bands) may cause some interference with the built-in compass, which could limit its accuracy.

2

Just the Tasks, Ma'am!

Learn about the 30 or so Apple Watch faces, as well as how to customize them and add complications to them. You also discover how to turn your Apple Watch into an alarm, stopwatch, and timer.

Explore how to use your Apple Watch to make and take phone calls, send and receive text messages (including Scribbles), and read and manage your email. You also learn how to send your actual heartbeat to a friend or loved one and use your watch as a Walkie-Talkie.

Turn your Apple Watch into a miniature media outlet by adding Notifications on topics you care about, such as the current weather, stock market performance, breaking news, live sports scores, and more. Learn about Dock to quickly access your favorite apps. Then, see how to keep track of appointments with the Calendar app and how to use your Apple Watch as a GPS via the Maps app.

» **Understanding customizations and complications**

» **Switching between watch faces for Apple Watch**

» **Accessing the time within any app**

» **Using the World Clock app**

» **Mastering the Stopwatch, Timer, and Alarm apps**

Chapter **4**

It's About Time: Learning How to Set Watch Faces, Alarms, Timers, and More

pple Watch is a watch after all, so chances are you'll use it a lot to tell time. But unlike a traditional analog watch or even most digital watches, Apple Watch lets you choose the face you want. This way, you can select what you'd like to see and how you'd like the information displayed.

This chapter looks at the multiple watch faces you can choose from and customize so you can truly make Apple Watch your own. You'll also see the different ways to access the watch face on Apple Watch. You then see other time-related apps pre-installed by Apple, including the World Clock, Stopwatch, Timer, and Alarm apps.

Looking at the Built-In Watch Faces

Why be stuck with only one watch face when you can have several to choose from?

That's one of the reasons people like a smartwatch. Because you've got a screen that can show virtually anything on it, you can go with a classic analog face (yes, with moving hands!), a digital watch face (just numbers), a hybrid of the two, or even one with animation on it. Some faces give you a ton of information, so you can get caught up with a single glance at your wrist.

It's a breeze to change these watch faces whenever you like — see the "Choosing from the Various Watch Faces" section later in this chapter — and without having to open up the Apple Watch app on your iPhone. By default, your watch face is the one called *Infograph Modular* (for Apple Watch Series 5). With this default, you get other onscreen information — called *complications*, which you can add or remove, like date, activity, music, breathe, and so on.

TECHNICAL STUFF

You might be swept away by the many watch faces you can choose from and customize. Remember that to you need to make sure your software is up to date to see the latest set of watch faces. Whatever you do, "look for software updates" on your Apple Watch, and remember that the "set of watch faces might differ from what you see on your Apple Watch." (See the Apple Watch User Guide, http://help.apple.com/watch, for more information.) Not all watch faces are available in all regions, and some might only work with newer versions of the watch (such as the Radio complication appearing only on Apple Watch Series 3 and later).

Plus, in the Apple Watch App Store, you can download many more watch faces — in the form of an app — to truly make Apple Watch uniquely yours (such as a virtual Cuckoo Clock app). But to get you going, Apple has installed a few good (and customizable) built-in watch faces.

TIP

With watchOS 6 — the latest operating system update that debuted in the fall of 2019 — you now have new ways to mark time. The Taptic Engine (that slight vibration on your wrist) can silently tap the hour on your wrist, if you like, so you know the time even without looking at your watch! Or you can set a chime to ring every hour on the hour (like a grandfather clock). And when you hold two fingers on your watch face, it tells you the time out loud (this is on by default). To add an hourly "tap" or chime, launch the Settings app on your Apple Watch, select Sounds & Haptics, and tap My Watch at the bottom of the screen, followed by Notifications.

Activity (analog and digital)

With color-coded activity levels — Move, Exercise, and Stand — these watch faces show your daily activity progress superimposed over a traditional analog clock or beside a larger digital clock. With the analog Activity face, you can choose to view your activity rings in the familiar stacked design or as small subdials on the face. Each option has many available customizable complications to place around the screen.

Astronomy

Apple says it worked with astrophysicists to create this visually striking watch face, shown in Figure 4-1. It shows the time and date at the top of the screen, but you can turn the Digital Crown button to view Earth's rotation, the moon phases, and even the entire solar system — all accurately displayed in time. Consider it a throwback to the oldest way to tell time: with stars, planets, and our moon.

FIGURE 4-1:
The Astronomy watch face.

Breathe

This watch face, shown in Figure 4-2, was designed to encourage relaxation and to breathe mindfully. At any time in the day, just raise your wrist and follow the rhythm and motion of the Breathe face. You have three styles to choose from: Classic, Calm, and Focus.

FIGURE 4-2:
Namaste! Here
are a few of
the Breathe
watch face
options in
the Apple
Watch app on
iPhone.

Plus you can add multiple complications to see information in the corners, including: Activity · Alarm · AQI (Air Quality Index) · Audiobooks · Battery · Breathe · Calculator · Calendar · Cellular · Cycle Tracking · Date · ECG · Find My Friends · Heart Rate · Home · Mail · Maps · Messages · Moon Phase · Music · News · Noise · Phone · Podcasts · Radio · Rain · Reminders · Remote · Stocks · Stopwatch · Sunrise/Sunset · Timer · UV Index · Voice Memos · Walkie-Talkie · Weather · Weather Conditions · Wind · Workout · World Clock

California

Debuting in WatchOS 6, this is the first Apple Watch face with a California dial — a mix of Roman and Arabic numerals — that you can further customize (all Roman, Arabic, Arbic-Indic, or Devanagari). You can select the background color, hands, and complications. See Figure 4-3.

FIGURE 4-3:
A few options with the California watch face.

Chronograph

Resembling an analog stopwatch, the Chronograph watch face has one main analog clock — with hour, minute, and second hands — but also two additional smaller hands: one for total time and a second for lap times, as shown in Figure 4-4. These secondary faces are very customizable too. You can also choose to place additional information in each of the four corners, which is explained in greater depth later in this chapter in the "Differentiating Between Customizations and Complications" section.

FIGURE 4-4:
The Chronograph watch face.

Color

By twisting the Digital Crown button, you can choose a watch face color that suits your outfit, style, or mood. This classic analog face can be as minimalist or as busy as you like based on the number of complications you select. See Figure 4-5 for the Color watch face.

FIGURE 4-5:
The Color
watch face.

Explorer

Quite simply, you can find the Explorer watch analog face (Figure 4-6) only on Apple Watch with cellular. It prominently features green dots, which indicate cellular signal strength in your area. You can customize the color hands (red, red/white, red/gray), choose from four different styles, and add optional complications, such as Activity, Radio, Walkie-Talkie, News, and Moon Phase.

Fire and Water

This watch face (Figure 4-7) brings two earthly elements to Apple Watch! Apple says each animated film was shot in a custom model allowing fire and water to define the edges of the face and interact with the dial. This watch face animates whenever you raise your wrist or tap the display. You can choose Fire, Water, or

Fire and Water. With Apple Watch Series 4 and later, you can choose full screen or circular. There are more than a dozen complications too.

FIGURE 4-6:
On Apple Watch with GPS + Cellular, the Explorer watch face has green dots that show you the strength of the cellular connection near you.

FIGURE 4-7:
Choose between Fire, Water, or Fire and Water together with the animated Fire and Water watch face.

Gradient

As shown in Figure 4-8, the Gradient watch face is simple and elegant, which lets you choose from three styles — of full-screen or circular gradients — each of which move with the time in different ways. You can add complications, too, to the top left, top right, bottom left, bottom right, or top middle of the watch face.

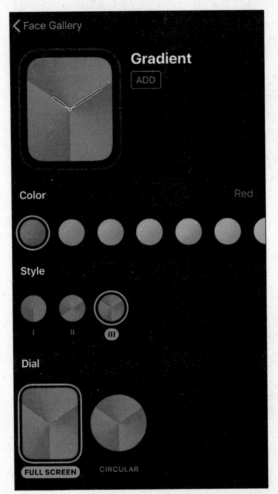

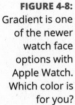

FIGURE 4-8: Gradient is one of the newer watch face options with Apple Watch. Which color is for you?

Infograph, Infograph Modular

Available on Apple Watch Series 4 and newer, the Infograph watch face (Figure 4-9) features up to eight rich, full-color complications and subdials, whereas the Infograph Modular face has up to six rich, full-color complications. You can choose from one of many dozens of colors for the outside lining of the face.

FIGURE 4-9:
Infograph (left)
and Infograph
Modular (right)
watch faces
offer a ton of
customizable
information
you can see at
a glance.

Kaleidoscope

As you might surmise, with a watch face called Kaleidoscope (shown in Figure 4-10), you can select a photo to create a watch face with evolving patterns of shapes and colors; simply turn the Digital Crown to change the pattern. You can choose a custom photo, or select from one of the many listed in the Apple Watch app for iPhone (Mirror, Flower, Graphic, and more). There are also multiple Styles to choose from and customizable complications.

FIGURE 4-10:
Trippy, man!
The Kaleido-
scope watch
face for Apple
Watch features
evolving
patterns and
shapes, with
lots of different
options
available too.

Liquid Metal

With four different colors (and two different styles) to choose from, the Liquid Metal watch face (Figure 4-11) animates whenever you raise your wrist or tap the display. Seriously, this looks super cool! With Apple Watch Series 4 and newer, you can select either a full screen or circular style.

FIGURE 4-11: Liquid Metal watch face options add a high-tech look to your Apple Watch.

Available complications: Activity · Alarm · AQI (Air Quality Index) · Audiobooks · Battery · Breathe · Calculator · Calendar · Cellular · Cycle Tracking · Date · ECG · Find My Friends · Heart Rate · Home · Mail · Maps · Messages · Moon Phase · Music · News · Noise · Phone · Podcasts · Radio · Rain · Reminders · Remote · Stocks · Stopwatch · Sunrise/Sunset · Timer · UV Index · Voice Memos · Walkie-Talkie · Weather · Weather Conditions · Wind · Workout · World Clock

Mickey Mouse, Minnie Mouse

The classic Mickey Mouse and Minnie Mouse analog watches — shown in Figure 4-12 — have been reinvented for Apple Watch. Mickey's or Minnie's arms move around the dial as they point to the correct hour and minute while his or her foot taps every second. You can choose from various customizations. In addition, you can choose whether your character speaks the time! To do so, open the Apple Watch app on your iPhone, tap My Watch, tap Sounds & Haptics, then tap Tap to Speak Time. You won't hear the voice if your Apple Watch is in Silent mode.

FIGURE 4-12:
The Mickey
Mouse
watch face.

Modular/Modular Compact

As the name suggests, this digital watch face allows for the most number of complications out of all watch faces, which gives you a ton of extra information at a glance. The interface can be as clean or cluttered as you see fit. With Modular Compact, shown in Figure 4-13, you have even more flexibility when adding complications, colors, and choosing between analog and digital time.

OH, MICKEY, YOU'RE SO FINE

This isn't the first time Apple has put Mickey Mouse on its gadgets. Apple launched a new iPod nano back in the fall of 2011 — and with it came 16 new clock faces, including ones with beloved Disney characters Mickey Mouse and Minnie Mouse.

Did you know the original Mickey Mouse watch debuted back in 1933 and became an instant hit? In fact, it saved the cash-strapped Ingersoll company from bankruptcy (like with many others companies, the Great Depression hurt Ingersoll's business considerably). The Mickey Mouse watch originally sold for $2.98 at Macy's and $2.69 at Sears, Roebuck & Company.

FIGURE 4-13:
The new
Modular
Compact
watch face.

Motion

Apple Watch fans are going to love this one. This watch face displays a different animated image every time you raise your wrist. Based on the theme you choose, raise your wrist and you might see a butterfly slowly fluttering its wings — as shown in Figure 4-14 — or a flower blooming. There are many colors and complications to choose from too.

TECHNICAL STUFF

Apple always goes above and beyond. Some of these objects for the Motion watch face were video-recorded, such as the jellyfish (at 300 frames per second), whereas others — such as the blooming flowers — were created using stop-motion, time-lapse photos. Apple says a single flower took more than 285 hours and 24,000 shots to photograph.

Numerals Duo/Numerals Mono

Fancy Schmancy! The Numerals Duo watch face displays the digital time in a large font — designed exclusively for Apple Watch, says Apple — with many customizable elements. The Numerals Mono face offers a digital and analog hybrid, with different colors to choose from. Why not change the colors and styles to best match your outfit? See Figure 4-15.

FIGURE 4-14:
The Motion
watch face.

FIGURE 4-15:
An example of
the Numerals
Duo and
Numerals
Mono watch
faces.

Available complications: Activity · Alarm · AQI (Air Quality Index) · Audiobooks · Battery · Breathe · Calculator · Calendar · Cellular · Cycle Tracking · Date · ECG · Find My Friends · Heart Rate · Home · Mail · Maps · Messages · Moon Phase · Music · News · Noise · Phone · Podcasts · Radio · Rain · Reminders · Remote · Stocks · Stopwatch · Sunrise/Sunset · Timer · UV Index · Voice Memos · Walkie-Talkie · Weather · Weather Conditions · Wind · Workout · World Clock

Photos

What better way to personalize your watch than by having a photo on it as your clock? With the aptly named Photos face (Figure 4-16), you can display a photo (or multiple photos) from your iPhone or iCloud gallery. In fact, set things up so a new photo appears every time you raise your wrist or tap your display. Choose from a synced album, recent Memories, or up to 10 custom photos. To add more photos, firmly press on the watch face, swipe all the way to the right, tap the New button (+), then tap Photos. Or, while in the Photos app on your Apple Watch, press any photo and then tap Create Watch Face.

FIGURE 4-16:
One of the most fun watch face options is Photos because you can see a different photo every time you lift your wrist!

Pride

Inspired by the rainbow flag — a symbol of LGBT pride (lesbian, gay, bisexual, and transgender) — this watch face offers multiple colored ribbons that move if you tap the display (see Figure 4-17).

Available complications: Activity · Alarm · AQI (Air Quality Index) · Audiobooks · Battery · Breathe · Calculator · Calendar · Cellular · Cycle Tracking · Date · ECG · Find My Friends · Heart Rate · Home · Mail · Maps · Messages · Moon Phase · Music · News · Noise · Phone · Podcasts · Radio · Rain · Reminders · Remote · Stocks · Stopwatch · Sunrise/Sunset · Timer · UV Index · Voice Memos · Walkie-Talkie · Weather · Weather Conditions · Wind · Workout · World Clock

Simple

The most minimalist of all Apple Watch faces, Simple — as the name implies — offers a straightforward yet elegant face with analog hands for minute, hour, and second, as shown in Figure 4-18. A single number tells you the day of the month. But as with all other watch faces, you can adjust the amount of detail with the Digital Crown button.

Siri

Your favorite personal assistant can now be part of your clock face. With the Siri watch face, you're A.I. pal takes a look at your day and displays information that's timely, relevant, and helpful (as shown in Figure 4-19). This could be the traffic on your way home from work, when your next calendar appointment is, or the sports score of your favorite team. Tap for a deeper dive or twist the Digital Crown to scroll through your day.

FIGURE 4-18:
The Simple
watch face.

FIGURE 4-19:
As seen here
in the Face
Gallery section
of the Apple
Watch app for
iPhone, you
can select a
Siri watch face
and customize
what info
your personal
assistant
shows you.

Solar Dial

Along with showing you the time in analog or digital format, Solar Dial features a 24-hour circular dial that tracks the sun's position in the sky above, as shown in Figure 4-20. Twist the Digital Crown to trace the sun's arc across the day.

FIGURE 4-20:
The Solar Dial
watch face.

TimeLapse

Who doesn't love time-lapse videos? Now you can see one — or a cityscape or natural setting — every time you raise your wrist to glance at the time (shown in Figure 4-21). You can choose which Apple "TimeLapse" video to see: Mack Lake (in California), New York, Hong Kong, London, Paris, or Shanghai.

Available complications: Activity · Alarm · AQI (Air Quality Index) · Audiobooks · Battery · Breathe · Calculator · Calendar · Cellular · Cycle Tracking · Date · ECG · Find My Friends · Heart Rate · Home · Mail · Maps · Messages · Moon Phase · Music · News · Noise · Phone · Podcasts · Radio · Rain · Reminders · Remote · Stocks · Stopwatch · Sunrise/Sunset · Timer · UV Index · Voice Memos · Walkie-Talkie · Weather · Weather Conditions · Wind · Workout · World Clock

FIGURE 4-21:
The TimeLapse
watch face.

Toy Story

If animated films starring *Toy Story* characters are more your pace than classic Mickey Mouse movies, Apple Watch has you covered just the same (see Figure 4-22). The Toy Story watch face lets you see animated *Toy Story* characters — Woody, Buzz, Jessie, and Toy Box — whenever you raise your wrist to look at the time. There are a dozen or so available complications, as well.

FIGURE 4-22:
To infinity and beyond! A look at some of the Toy Story watch face options.

Utility

The most straightforward and practical face out of the bunch, Utility shows you a classic analog watch face but with plenty of space in the corners for extra information, such as the world clock, the timer, or a calendar appointment, as shown in Figure 4-23.

FIGURE 4-23:
The Utility
watch face.

Vapor

The second-to-last watch face for Apple Watch is Vapor (Figure 4-24), which animates . . . er, vapor, whenever you raise your wrist or tap the display. As for how it was created, Apple says each of the four films was shot at thousands of frames per second as a mortar fired colors into a custom chamber. Cool! Choose the color and style to suit your tastes. There are multiple complications to choose from too.

X-Large

As you might expect with a name like X-Large, this watch face shows a very large digital clock with the hour on the top of the screen and the minutes on the bottom, as shown in Figure 4-25. Although the background is black, you can adjust the color to your liking — perhaps to match an outfit you're wearing.

FIGURE 4-24:
Mysterious and sophisticated: the Vapor watch face for Apple Watch.

FIGURE 4-25:
The X-Large watch face.

CAN APPLE MAKE THE WATCH RELEVANT AGAIN?

Smartwatch naysayers claim that many of us have moved on from the watch and now rely on our phones to tell time. Therefore, even a so-called "smart" watch won't appeal to that many people. I would argue that having the time on your wrist is certainly more convenient than pulling out that phone from your back pocket or purse.

"Time" will tell if Apple and other smartwatch makers can make the watch relevant again. I'd argue the wrist is indeed coveted real estate to add some technology to; telling time is just the beginning.

That said, Apple's CEO Tim Cook made it very clear Apple Watch should not only be a functional timepiece but also an accurate one.

Choosing from the Various Watch Faces

One of the first things you might do with Apple Watch is to choose one of the many, many watch faces provided and then customize it to your liking — perhaps with complications. And it's super easy to change faces and make those optional changes on the watch itself.

The first step is to load a few desirable faces onto the watch, which you can handle wirelessly via your iPhone! To load watch faces, follow these easy instructions:

1. **Open the Apple Watch app on your iPhone.**

2. **Tap where it says Face Gallery at the bottom of the screen.**

3. **Scroll up and down the screen, as shown in Figure 4-26, to see multiple watch faces to choose from, as well as some options in style and complications.**

4. **If any watch face looks interesting to you, tap the Add button and your iPhone beams the watch face over to Apple Watch, like magic.**

FIGURE 4-26:
The Face
Gallery section
of the Apple
Watch app (on
iPhone).

That's it!

Now you'll have a selection of clock faces to choose from on the Apple Watch itself. To select a specific watch face for your Apple Watch, follow these steps:

1. **When viewing the default (Chronograph) watch face, press and hold the screen.**

   This enables Force Touch and launches the Faces gallery.

2. **Swipe left or right to select a clock face you like, and then tap the center of the watch face you want to use this time.**

3. **Tap Customize near the bottom of the screen to personalize the face.**

   The small white dots on the top of the screen tell you how many different customization screens are available for this face (in Figure 4-27, only one white dot appears). You can change whatever is in the green outline.

FIGURE 4-27:
Whatever is highlighted in green can be customized by twisting the Digital Crown button.

4. **Twist the Digital Crown button to customize the screen.**

   When you like what you see — for example, changing the color of the hands or text or adding a second hand — swipe to the left to go to the next customization screen. I cover customizations in greater depth in the "Differentiating Between Customizations and Complications" section.

5. **Twist the Digital Crown button to make your selection once again.**

   Repeat this step until you've gone through all the customization screens — perhaps choosing color and extra time information. Typically, the last customization page is for complications.

6. **Tap the green areas that you can add to your watch screen, as shown in the left image of Figure 4-28, and then twist the Digital Crown button to select what you're happy with.**

   Repeat the process by tapping the other areas of the face you want to change, which vary by which watch face you go with. Remember, you won't have the same options for all watch faces.

7. **Press the Digital Crown button when you're done customizing your watch face.**

   This confirms you've finished with your options and are ready to see your customized watch face.

8. **Tap the center of the screen to confirm your changes.**

   After you've set your watch face, don't forget you can tap on each of the complications — such as weather or stock quotes, as shown in the right image of Figure 4-28 — which are discussed in further detail in the "Differentiating Between Customizations and Complications" section.

FIGURE 4-28: If you don't like your customization choices, you can go in at any time and tweak them (left). After you customize, this is the watch face you might see when you raise your wrist.

That's it! That wasn't so difficult, right?

Differentiating Between Customizations and Complications

It's important to understand that all the watch faces support various tweaks you can make to help your Apple Watch feel even more personal. Apple divides these personalized changes into two main categories: customizations and complications.

Customizations

Each watch face includes customizations you can perform, such as changing the watch to show additional information (such as specific minute, second, or millisecond detail) and changing the color of the watch hands (perhaps to match your outfit). You can make changes for each face, but what you can change varies by face style. See Figures 4-29 for customization examples.

Complications

The word *complications* is a horological term that refers to any clock feature that goes beyond the display of hours and minutes.

When customizing your watch's face, complication options mean you can add additional information to the screen, usually in up to four corners or at the bottom part of the screen. This may include upcoming calendar alerts, an alarm, the current moon phase, the weather, the day's sunrise and sunset times, a timer or stopwatch, your current activity, the world clock, and stock quotes. To customize your display, simply touch highlighted parts of the watch face to make your selection. Be aware that some watch faces offer more room for complications than others. When you tap the piece of information provided, such as the weather, Apple Watch opens the corresponding app for a deeper dive. See Figures 4-30 and 4-31 for a few examples.

FIGURE 4-30: What can you choose to place on your watch face? A lot, as you can see here.

FIGURE 4-31: An example of a kind of complication you can add to a watch face.

TIP

You can still tell the time on Apple Watch if the battery is low. Your watch automatically goes into the Power Reserve mode when your battery drops to a certain percentage, or you can activate this mode manually by opening the Settings app on Apple Watch. You should still be able to see the time for up to 48 hours, according to Apple.

Looking closer at a few complications

You can add one of 30-odd complications to your Apple Watch face (if it's supported on that specific face):

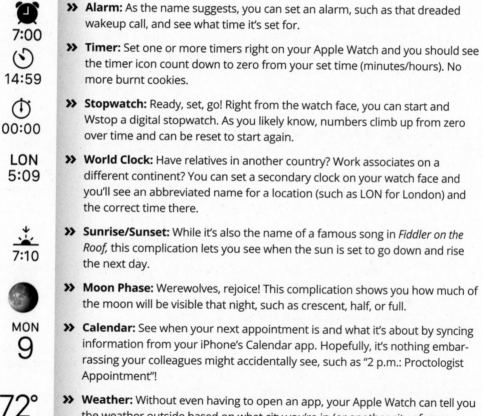

- >> **Alarm:** As the name suggests, you can set an alarm, such as that dreaded wakeup call, and see what time it's set for.

- >> **Timer:** Set one or more timers right on your Apple Watch and you should see the timer icon count down to zero from your set time (minutes/hours). No more burnt cookies.

- >> **Stopwatch:** Ready, set, go! Right from the watch face, you can start and Wstop a digital stopwatch. As you likely know, numbers climb up from zero over time and can be reset to start again.

- >> **World Clock:** Have relatives in another country? Work associates on a different continent? You can set a secondary clock on your watch face and you'll see an abbreviated name for a location (such as LON for London) and the correct time there.

- >> **Sunrise/Sunset:** While it's also the name of a famous song in *Fiddler on the Roof,* this complication lets you see when the sun is set to go down and rise the next day.

- >> **Moon Phase:** Werewolves, rejoice! This complication shows you how much of the moon will be visible that night, such as crescent, half, or full.

- >> **Calendar:** See when your next appointment is and what it's about by syncing information from your iPhone's Calendar app. Hopefully, it's nothing embarrassing your colleagues might accidentally see, such as "2 p.m.: Proctologist Appointment"!

- >> **Weather:** Without even having to open an app, your Apple Watch can tell you the weather outside based on what city you're in (or another city of your choosing). You can choose to see the information in Fahrenheit or Celsius. Ideal for Canadians, eh?

- >> **Stocks:** Should you celebrate or bite your nails? The Stocks complication shows you real-time quotes based on the publicly traded companies that matter to you.

>> **Activity:** Get a quick idea of how active you've been throughout the day. AThe three rings show your actual level of physical activity and how much of your daily goal you've accomplished. Lazy types might be tempted to throw the watch out, but it won't berate you.

>> **AQI, UVI:** Glance at your local AQI (Air Quality Index) or UVI (Ultraviolet Index) as complications at the bottom right and bottom left of the Sun Dial watch face. With both of these indices, the higher the number, the more risk of harm from poor air quality and unprotected sun exposure, respectively.

Here are a bunch of other complications available, depending on the watch face you've selected:

Audiobooks · Battery · Breathe · Calculator · Cellular · Compass · Cycle Tracking · Date · ECG · Find My Friends · Heart Rate · Home · Mail · Maps · Messages · Music · News · Noise · Phone · Podcasts · Radio · Rain · Reminders · Remote · Stopwatch · Voice Memos · Walkie-Talkie · Weather Conditions · Wind · Workout

Accessing Time on Apple Watch

If you've mastered how to select a watch face and customize it, you might want to know how to access the clock on your Apple Watch. It's probably the screen you're going to see the most after all, right?

It's actually quite easy, even if you're in another app.

By default, Apple Watch shows you the clock screen when you raise your wrist. Therefore, you need not do anything if you like this setup. The screen stays on for about four to six seconds — whether you look at it or not — and then goes back to sleep again (presumably to save on power).

If you tap the screen or press the Digital Crown button or the side button, however, the screen stays on for about 15 to 17 seconds before fading to black.

Want to see something other than the time when you raise your wrist? No problem. Grab your iPhone (because you need it to change the default setting to another option) and then follow these steps:

1. **From the Apple Watch app on your iPhone, choose My Watch ⇨ Settings ⇨ General.**

   Near the bottom of the screen, you should see the words *Wake Screen*, as shown in Figure 4-32.

2. If you like seeing the time when you raise your wrist, do nothing.

If not, make some changes here by selecting one of the other options listed, such as seeing the last app you were in.

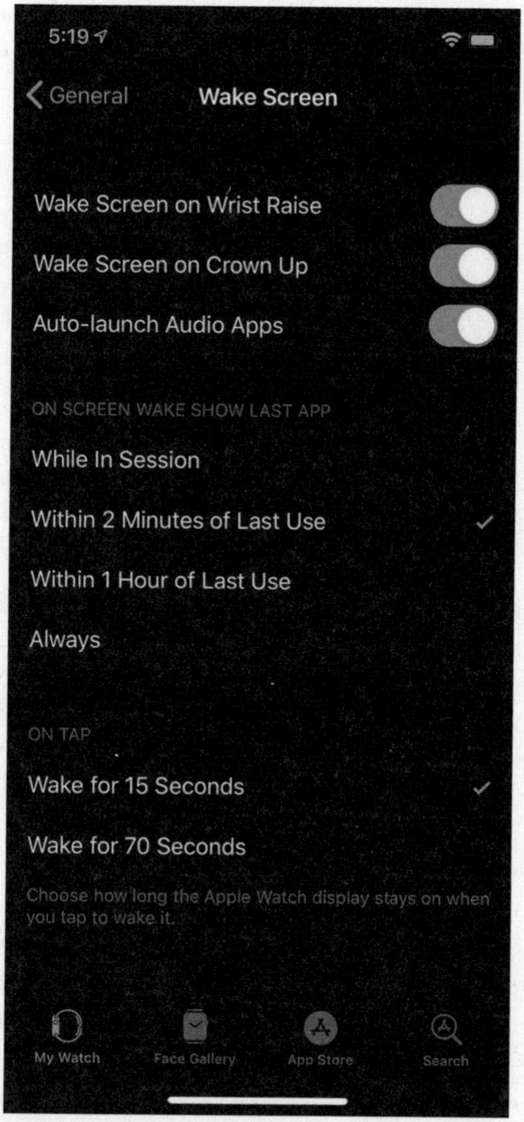

5:19

‹ General   **Wake Screen**

Wake Screen on Wrist Raise

Wake Screen on Crown Up

Auto-launch Audio Apps

ON SCREEN WAKE SHOW LAST APP

While In Session

Within 2 Minutes of Last Use ✓

Within 1 Hour of Last Use

Always

ON TAP

Wake for 15 Seconds ✓

Wake for 70 Seconds

Choose how long the Apple Watch display stays on when you tap to wake it.

My Watch   Face Gallery   App Store   Search

FIGURE 4-32:
The Apple Watch app on iPhone. Here, you can enable and disable various Apple Watch functions.

You can also easily see the time whenever you're in any app (as further described in Chapter 3). Simply double-press the Digital Crown button and you immediately see the watch face appear on your screen. Now double-press the Digital Crown button again and you return to the last app you were in.

You have a third way to get the time on your Apple Watch, and you don't even need to look at your wrist at all. Can you guess what it is? Give up? You can ask Siri what time it is. Press and hold the Digital Crown button and ask "What time is it?" or simply say "Time." Alternatively, you can ask "What time is it in _____ (and name a city around the world)"? You can also raise your wrist and say "Hey, Siri, what time is it?" See Chapter 7 for more on using Siri to help you complete tasks with your Apple Watch.

Accessing World Time

Some people like to know what time it is in another part of the globe. You know, in case you want to Skype or Facetime with someone and you're not sure if it's the middle of the night where that person is.

Whether it's for personal or professional reasons, a world clock could be a handy thing to have, and your Apple Watch can help you with that.

Earlier in this chapter — in the "Differentiating Between Customizations and Complications" section — I discuss adding a world clock as a complication to an existing watch face, but it's also something you can view on your own.

To use the World Clock app on your Apple Watch, follow these steps:

1. **Press the Digital Crown button to go to the Home screen.**

   Regardless of the app you're in, you should see the screen with all the small icons on it after you press the Digital Crown button. If you don't see what you want at first glance, swipe your finger around to view other bubble-shaped icons.

2. **Tap the World Clock app.**

   The app launches full screen and you should see the time in another city. If you haven't added another city yet, Apple Watch might say "No World Clocks," such as it does on an iPhone.

3. **If you have more than one city selected, you can swipe left or right to navigate between them.**

   The top right of the screen is your local time. But then you should see the remote city highlighted by an orange dot on a map, the name of the city, the time there, the time zone, and sunrise and sunset information, as shown in Figure 4-33.

4. **If you don't have any locations installed or if you'd like to add more (or remove one), open the Settings tab on your iPhone's Apple Watch app (under General).**

When you've made your selection, you can just exit the app.

FIGURE 4-33:
When you launch the World Clock app, your screen should look similar to this.

Of course, you can also ask Siri for this information, such as "What time is it in Warsaw?" or "What time will the sun rise in Tokyo?" See Chapter 7 for more tasks Siri can help you with.

Taking Control: Alarms, Stopwatches, and Timers — Oh My!

Alarms, timers, and stopwatches are great tools to have with you. And because you're wearing Apple Watch on your wrist, you have access to them wherever life takes you.

Alarms

Apple Watch lets you easily set an alarm, such as a 7 a.m. wakeup call, viewed in either an analog or digital display. To change between digital and analog, press firmly on the screen and then tap Customize. Swipe left until you see the alarm face you like.

To set up a new alarm on Apple Watch, follow these steps:

1. **Press the Digital Crown button to go to the Home screen.**

2. **Tap the Alarm app.**

   This launches the Alarm app. From this screen, you can view, manage, and edit multiple alarms with your fingertip.

   To set an alarm with Siri, press and hold the Digital Crown button to activate Siri, or simply raise your wrist and say "Hey, Siri," followed by "Set an alarm for ___ (date and time)." You can also say, "Wake me up at ___ (time of day)" or even "Wake me up in ___ (minutes or hours)."

   Siri confirms the time and shows it to you on the watch's screen, as shown in analog mode in Figure 4-34 (left). It's totally fine to have multiple alarms.

3. **To set your alarm time, twist the Digital Crown button to adjust the hours and minutes and then tap Set.**

   The right image in Figure 4-34 shows an alarm setting.

   Your alarm is now set. You can uncheck ones you no longer need. You can also press and hold the Apple Watch screen in the Alarm app to bring up options, including alarm repeats (such as for weekdays). When the alarm goes off, you can tap Snooze or Dismiss.

TIP

Your iPhone alarms can also be synced with your Apple Watch. This happens automatically by default, but you can also go into your Apple Watch app on an iPhone and tap My Watch and then Settings to disable this feature.

Stopwatches

You don't need to be an Olympic runner to appreciate a stopwatch, which measures the ascending passage of time. Whatever the reason you'd like to know how much time has elapsed, the Stopwatch app is what you need — and it's fully customizable too.

That is, the Stopwatch app on Apple Watch lets you see information in a digital, analog, or hybrid view, or even in a graph that shows a real-time average of your lap times. See Figure 4-35 for a look at the hybrid view.

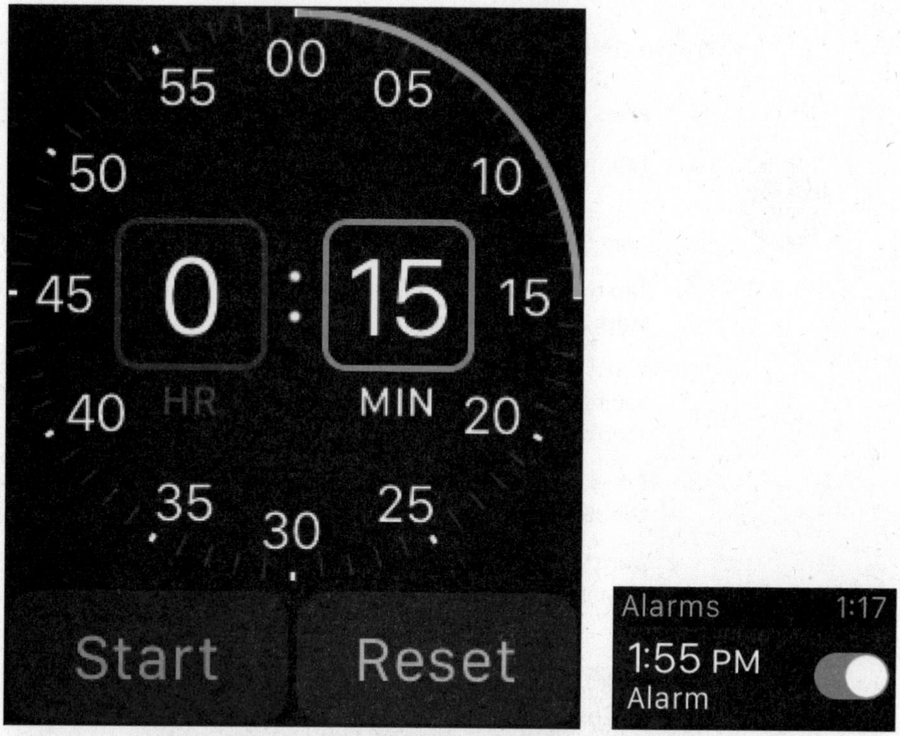

FIGURE 4-34: The analog mode of the Alarm screen. Set your desired time using the Digital Crown button (left). It's a cinch to adjust your alarm time — even if you're half asleep and fumbling to set a wakeup call (right).

FIGURE 4-35: You can choose a hybrid of analog and digital, as shown here.

To use the Stopwatch app on Apple Watch, follow these steps:

1. **Press the Digital Crown button to go to the Home screen.**

2. **Tap the Stopwatch app.**

   This launches the Stopwatch app. You can also raise your wrist and then say "Hey, Siri, Stopwatch," or press and hold the Digital Crown button to initiate Siri.

3. **Tap the green Start button in the lower left, as shown in Figure 4-36, to start the stopwatch.**

   Whether you're in analog, digital, or hybrid view (which you can change in the Settings area of the iPhone's Apple Watch app, as discussed in Chapter 11), you should see the time scroll by.

4. **Press the red Stop button in the lower-right corner of the screen to stop the stopwatch.**

   Don't worry if you accidentally close the app without taking note of your time because it's still there when you open up the app again.

FIGURE 4-36: The Stopwatch app in action.

5. **Tap the Lap button in the lower right corner of the app if you want to see graphed averages of lap times.**

Joggers and runners might appreciate this added historical information, as shown in Figure 4-37.

FIGURE 4-37: The Stopwatch app offers a historical/ graphical look of your lap times.

Timers

The Timer app lets you keep track of events you want . . . well, timed. Whereas the Stopwatch app measures the ascending passage of time, the Timer app offers the descending passage of time from a preset starting point, such as 45 minutes.

As with the other time-related apps, you can choose to read a digital or analog countdown.

To use the Timer app on your Apple Watch, follow these steps:

1. **Press the Digital Crown button to go to the Home screen.**

2. **Tap the Timer app.**

This launches the Timer app. Or you can say "Hey, Siri, Timer" or press and hold the Digital Crown button to activate Siri and then say "Timer."

3. **Twist the Digital Crown button to select the starting time.**

You start with the hour setting, but press the Digital Crown button to go to the minute setting and then press the Digital Crown button again when you're done. Review what you selected for a start time, such as 30 minutes or an hour and 15 minutes.

4. **Tap the green Start button on the lower left side to initiate the timer.**

As the timer runs, an orange line moves around the dial in clockwise fashion to help give you a visual sense for how much time has passed (and how much is left). See Figure 4-38 for a look at the Timer app in all its glory.

5. **When you're done, press the Reset button in the lower-right corner of the Apple Watch screen.**

Pressing Reset rolls back the timer to all zeros, making it ready for next time. If you no longer need the timer, press the Digital Crown button.

FIGURE 4-38:
The analog and digital hybrid screen of the Timer app.

IN THIS CHAPTER

» **Accepting and placing calls on your Apple Watch**

» **Handing off calls to your iPhone**

» **Receiving and sending messages, including the Scribble feature**

» **Sending Animojis and Memojis, stickers, handwritten messages, voice clips, taps, heartbeats, and more**

» **Using the Walkie-Talkie feature**

» **Receiving and managing emails**

Chapter **5**

Keep in Touch: Using Apple Watch for Calls, Messages, Texts, Walkie-Talkie, Emails, and More

A pple Watch isn't used just for information — such as the time, the weather, and sports scores — but it's also ideal for keeping in touch with those who matter.

In other words, your smartwatch isn't just about knowledge; it's also about communication. That's precisely what this chapter is all about. On one hand, if you can pardon the pun, you've got the same familiar ways to connect as you do on your iPhone. Specifically, your Apple Watch can place and accept calls as well as send messages and emails. Although Apple Watch was designed for quick interactions, it supports many of the same chatting features as your smartphone.

In fact, if you own one of the Apple Watch models with GPS + Cellular and have set up a wireless plan with your phone provider, you don't even need your iPhone nearby to place or accept a call or text message! But it can also be used to reach out to others in new and unique ways, such as sending someone a tap or your heartbeat, which he or she will feel on his or her wrist (that is, if that person is wearing an Apple Watch, which he or she must to feel these sensations). You can also send a finger-drawn sketch to someone special.

Accepting and Placing a Call on Apple Watch

If you want to be like Dick Tracy and take calls on your wrist, Apple Watch lets you do just that. Or you can make a call by pressing the Digital Crown button and asking Siri to call someone.

You have two ways to do this, depending on which model Apple Watch you have:

>> **With Cellular connectivity:** Whether you initiate the call or accept it, as long as you have a cellular plan on a supported Apple Watch (GPS + Cellular model), you can chat through your smartwatch's microphone and hear the other person through the small speaker. When your watch connects to a cellular network, four green dots appear in the watch's Control Center.

>> **Without Cellular connectivity:** If you don't have an Apple Watch that supports cellular connectivity, you can still use your Apple Watch to chat with others. You handle this via Bluetooth technology with your nearby iPhone — up to a few dozen feet away — or even farther than that over Wi-Fi. As long as your iPhone is connected to the same wireless network — such as at home or at the office — your watch rings at the same time as your phone, and you can choose which one to answer (or not).

DIDYA KNOW?

You can set up Apple Pay on your Apple Watch — via your iPhone — and send someone virtual money through the Messages app. Learn more about it in Chapter 10.

Incoming calls

This is the easy one.

If a call comes in to your phone number, your Apple Watch rings just like your iPhone — unless you choose to disable that feature in the Apple Watch app on your iPhone, which you can learn how to do in Chapter 11.

Assuming you didn't mess with the default settings, you hear your ring tone emanate through your watch's speaker and see a screen pop up with the name of the person calling (or just a phone number if that person isn't in your Contacts). Tap the big green icon ("Answer" the call) in the bottom right of your watch's screen to answer the call.

If you don't want to answer the call, tap the big red ("Hang Up") icon in the lower left of your Apple Watch. Alternatively, you can swipe up from the bottom of your watch to send a preset message back to the person trying to reach you, such as "Call you later," "In a meeting," or "I'm driving." Just make sure that call is coming from a mobile phone or else the caller may not see it.

Figure 5-1 shows a couple preset messages you can send back if you're unable (or unwilling) to speak.

FIGURE 5-1:
You can respond to incoming calls by sending a customized reply in the Apple Watch app on your iPhone.

You can customize these preset replies by going into the Settings ⇨ Messages area of the Apple Watch app on your iPhone; see Chapter 11 for more on doing this. But as infomercial guru Ron Popeil once famously said: "But wait — there's more!"

> » **Transfer it.** You can transfer your call to your iPhone, a Bluetooth headset, or a car's speakerphone. See the "Handing Off a Call to Your iPhone or via Bluetooth" section for more on this.

> » **Stop your wrist from ringing.** Perhaps you're in a crowded elevator and you're getting dirty looks from someone. You can silence an incoming call by simply covering Apple Watch with your other hand. You can try to cover it with the same hand the watch is on, but that may prove a tad difficult.

> » **Mute the ring.** The Phone app screen on Apple Watch also lets you mute your microphone by tapping the microphone with a slash through it (in case you don't want other participants in a conference call to hear you sneeze!). At the top of the watch screen, you can also increase or decrease the volume coming through the watch's speaker.

Outgoing calls

While certainly not difficult, outgoing calls through your Apple Watch requires a little more work. And that's assuming you actually *want* to make a call through your wrist.

A few considerations: It may not be too comfortable to hold up your wrist for an extended period of time; the quality of the call won't be as good as on a phone; your conversations may be heard because it's a speakerphone (unless you're wearing a Bluetooth headset); and, oh, you might look a little silly too.

The Apple website also suggests you might not want to talk for long on Apple Watch anyway: "Use the built-in speaker and microphone for quick chats, or seamlessly transfer calls to your iPhone for longer conversations." With that in mind, you can place a call on your Apple Watch in a few ways:

> » **First approach:** Press the Digital Crown button to go to your Home screen and then tap the Phone icon. Go ahead and dial the number in the Keypad section, or pull up someone in Contacts, Recents, or Favorites (see Figure 5-2). Swipe down or twist the Digital Crown to access one more option: Voicemail.

> Figure 5-3 shows you the other options that appear when you pull up a contact in the Phone app. When you're in the Contacts part of the Phone app, you can see the icons here for placing a call, texting (SMS) someone, or sending an email. Under that is Notes, if you've ever filled out that section on your iPhone. Yes, remember your Contacts info is synced with the Contacts app on your iPhone!

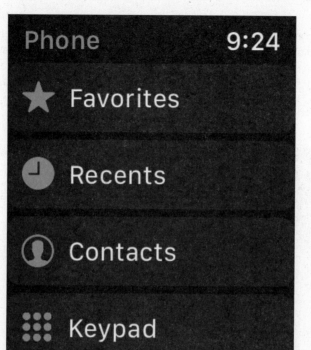

FIGURE 5-2:
When you open the Phone app, you see the main screen.

FIGURE 5-3:
Pulling up a contact in the Phone app reveals icons to call or text (left) as well as to email or view notes (right).

>> **Second approach:** Press the side button to bring up Dock and then twist the Digital Crown button or swipe up and down to get to the Phone section (if you've recently used the watch to make a call or set the Phone app as one of your favorite Dock apps). Tap the Call icon for a given contact, as shown in Figure 5-4.

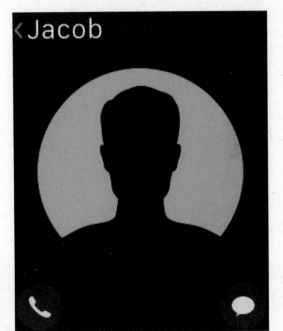

FIGURE 5-4:
Access the Phone section of Apple Watch by pressing the side button.

>> **Third approach:** Lift your wrist and say "Hey, Siri" into your watch, followed by "Call ____ (person's name)" (if that person is in your phone's Contacts list) or "Dial ____ (phone number)." Or you can press and hold the Digital Crown button to activate Siri. Or you can just say "Hey, Siri, make a call" and wait for it to ask you to name someone or provide a number, as shown in Figure 5-5.

Outgoing calls with FaceTime audio

You can't make video calls from your wrist — er, yet — but you can use FaceTime to place an audio call, if you like. You may be in a Wi-Fi hotspot and prefer to call over FaceTime (which uses data) over a cellular connection. Or the person you're calling may be on an iPad, and it's the only way to call them (see Figure 5-6). I whited out the phone numbers for this contact for privacy reasons, but you get the gist.

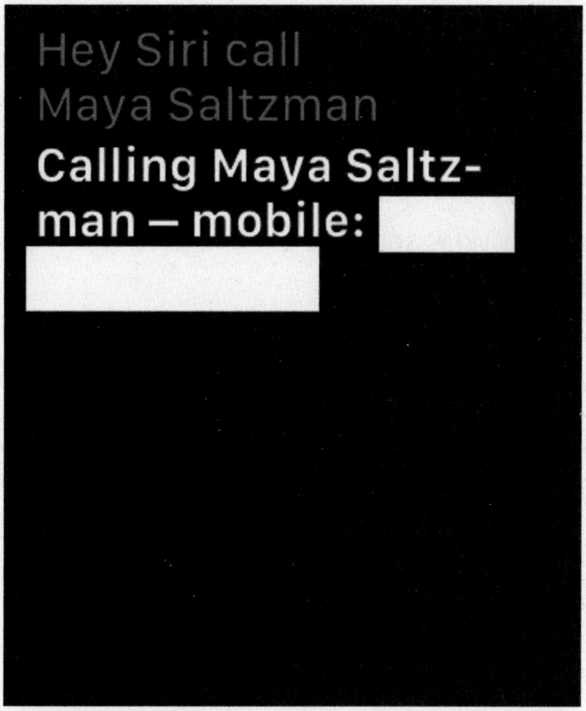

Hey Siri call Maya Saltzman

Calling Maya Saltz-man — mobile:

FIGURE 5-5:
The fastest way to make a call? Use Siri to dial a contact or phone number (actual phone number blocked out ☺).

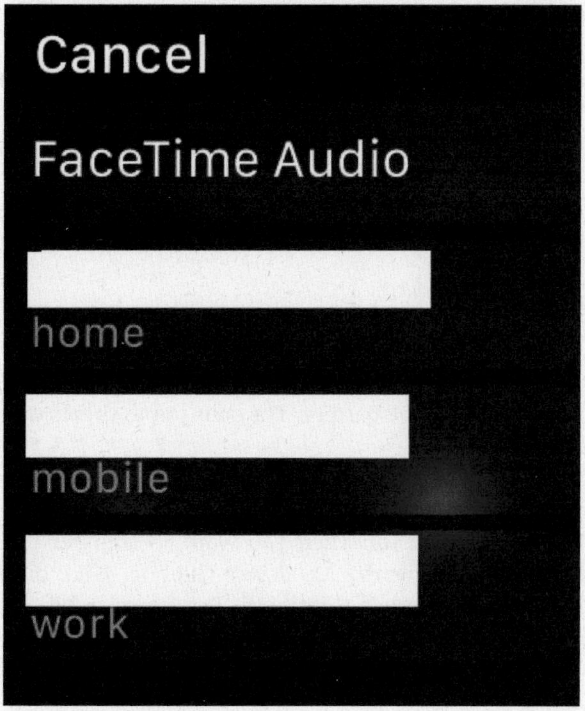

Cancel

FaceTime Audio

home

mobile

work

FIGURE 5-6:
It's super easy (and frankly, fun) to place a FaceTime audio call from your wrist. If the person you're calling has FaceTime, it will appear as an option.

To place a FaceTime call using the Phone app on your Apple Watch, follow these instructions:

1. **Open the Phone app on your Apple Watch (per preceding section).**

2. **Tap Contacts.**

   Tap on the contact you want to call.

3. **Tap the Phone icon.**

   You may see an option for FaceTime Audio (if your contact has FaceTime).

4. **Tap the option for FaceTime Audio.**

Alternatively, you can summon Siri and ask her to place a FaceTime audio call with someone too! Siri is the easiest. To recap, with examples: Raise your wrist or say "Siri," followed by:

>> "Call Mom"

>> "Dial 800 555 1212"

>> "Call Pete FaceTime audio"

TIP

If you ever want to turn the Apple Watch screen off or silence any sounds coming from it, simply cup your hand over the face and it should go dark and silent.

Not too confusing, right? You can always revert back to smoke signals or Morse code if all this technology is making your head spin!

Handing Off a Call to Your iPhone or via Bluetooth

You probably don't want to talk for long periods of time through your Apple Watch, if only because it eats up the battery. The solution to this is wirelessly handing off the call to your iPhone.

By the way, this "handoff" feature is available in other apps too, and not just for phone calls. It's ideal for when you want to transfer what you're doing to another compatible and nearby iOS device (iPhone, iPad, or iPod touch) or Mac computer. It's part of Apple's Continuity feature over Wi-Fi and includes such apps as Calendar, Reminders, Messages, Mail, Contacts, Maps, and more.

Handoff should already be enabled on your Apple Watch, but if it isn't for whatever reason, Figure 5-7 shows the Apple Watch app on your iPhone and where in the Settings you can enable the feature: My Watch ⇨ Settings ⇨ General. Simply flick the tab to green where it says Enable Handoff.

If an Apple Watch app can be "handed off" to your iPhone, you should see the Handoff option inside the watch app, and once selected, you can simply tap the icon in the lower-left corner of the iPhone screen to complete the handoff to your phone.

Apple says this about a handoff: "When this is on, your iPhone will pick up where you left off with apps on your Apple Watch. Apps that support this feature appear on the lower-left corner of your iPhone lock screen."

To use Bluetooth for handing off a call, when a call comes in, slide up from the bottom of the watch screen for a list of options, including handing a call off to another Bluetooth-enabled device — be it the iPhone itself, a hands-free Bluetooth headset, or perhaps a Bluetooth-enabled stereo in your vehicle. Doing this transfers the call to the desired device.

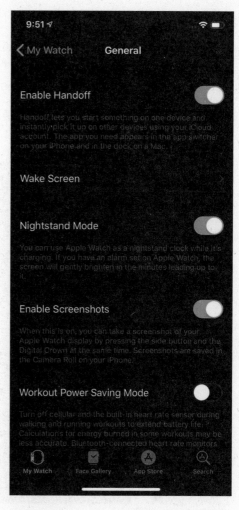

FIGURE 5-7:
You can tweak a number of options in the Apple Watch app on your iPhone, including transferring activities to other Apple products.

Making Apple Watch Calls over Wi-Fi

If your carrier supports Wi-Fi calling — as many today do — then you can place and receive calls over Wi-Fi rather than a cellular network. This is especially handy if you don't own an Apple Watch with GPS + Cellular because you can still use your Apple Watch to make calls, even if your iPhone isn't nearby. As long as you're in a Wi-Fi network, you're good to go.

To get going:

1. **On your iPhone, go to Settings ⇨ Phone, tap Wi-Fi Calling, then turn on both Wi-Fi Calling and Add Wi-Fi Calling for Other Devices.**

See Figure 5-8. If you can turn this on, your carrier supports it.

2. **Enable Wi-Fi calling on your Apple Watch by opening the Apple Watch app on your iPhone, tapping My Watch, then Phone, then turning on Wi-Fi Calls.**

FIGURE 5-8:
Enabling Wi-Fi calling on both your iPhone and Apple Watch lets you place or accept calls over a Wi-Fi network.

TIP

It's recommended to make emergency calls over a cellular connection rather than Wi-Fi because your location information is more accurate. To err on the side of caution, make sure your emergency address is up to date: on your iPhone, go to Settings ⇨ Phone ⇨ Wi-Fi Calling, then tap Update Emergency Address.

In the fall of 2019, Apple announced that cellular models of Apple Watch Series 5 can make international calls to emergency services regardless of where the device was originally purchased or if the cellular plan has been activated.

Ideal for world travelers, international emergency calling also works with fall detection, if enabled, to automatically place an emergency call if Apple Watch senses the user has taken a hard fall and remains motionless for about a minute (see Chapter 8).

Receiving and Sending Messages

Many millions of iPhone users send messages from one iPhone to another device, whether it's a standard text (SMS) message or through Apple's own iMessage service. That might be a bit confusing because the app itself is called *Messages*, but iMessage is part of it, along with regular texting. The main advantage to iMessage over texting is it's free for you and anyone writing over Wi-Fi, and it's unlimited, so you can type and upload media as much as you want.

Apple Watch also houses a Messages app — supporting iMessage and text messages — although unlike the iPhone, you don't have a keyboard for you to type words on the watch.

Although you can't type on Apple Watch, you can send messages through voice dictation (or send the audio clip instead), tap on preset responses based on the context of messages you've received, or send animated emojis (cute and customizable icons) or your location on a map.

Receiving and responding to messages

You can keep you iPhone tucked away yet still correspond with important people in your life. When a message comes in (via iMessage or a text message), your Apple Watch vibrates on your wrist (and dings) to let you know you have a new message waiting to be read. (You can disable tactile feedback and sound in the Settings area of the Apple Watch app on your iPhone, as discussed in Chapter 11.)

Acknowledging a message

To receive, reply, and initiate a message to someone via your Apple Watch, follow these steps:

1. **If you feel a pulse and hear a tone, raise your wrist to see the message.**

   As shown in Figure 5-9, you should see a screen showing whom the message is from, what he or she wrote, and perhaps an integrated image. You can scroll up and down by twisting the Digital Crown button if the message is longer than what's on the screen.

2. **To dismiss the message, lower your wrist or tap Dismiss at the bottom of the message.**

   Do this if you don't want to reply — or at least not right now. You can exit the Messages app and return later.

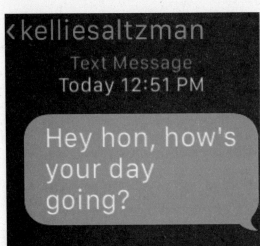

3. **If you want to reply to the message, tap Reply at the bottom of the screen and you should see several options (see Figure 5-10).**

   Apple Watch suggests some preset words to reply with based on the context of the conversation, along with some preset responses you can choose from (such as Not Sure, Can't Talk Now, Talk later?). Twist the Digital Crown button to see all the responses and then tap one you like. To create a custom response in the Apple Watch app on your iPhone, see Chapter 11.

Replying to a message

If you want to reply to a message with your voice, follow these steps in the Messages app:

1. **Tap the microphone icon to speak (dictate) your reply and have it transcribed into text or sent as an audio clip.**

   Speak clearly into your wrist and you should see the words typed as you say them.

2. **Tap Send in the top right of the screen.**

   You will see a preview of the words before you send them. You can't change the words if something isn't correct, so you have to tap Cancel in the top left and then speak again (perhaps slower and clearer!).

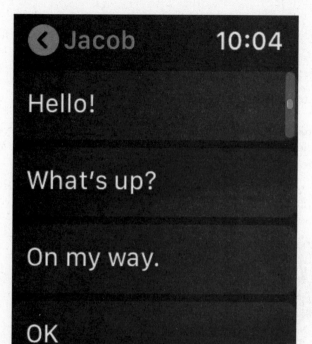

FIGURE 5-10:
You can use several preset replies to while sending someone a message.

TIP

If you don't see an option to send your message as a voice clip, go to Settings ⇨ Messages on the Apple Watch app on your iPhone (Figure 5-11) and select whether you want your watch to transcribe your voice into text, send it as a voice recording, or give you the option on your watch after you speak into it.

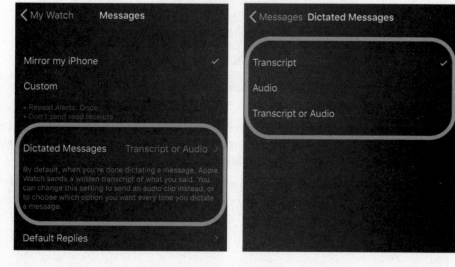

FIGURE 5-11:
Want to use your voice to reply to messages? The Messages area on the Apple Watch app (left) allows you to either send dictated text or a voice recording (right).

See the section "Responding with an audio clip" for more on sending an audio clip.

Responding with emojis

To reply to a message with an emoji, follow these steps:

1. **Instead of replying with a preset message, tap the smiley face on the Apple Watch screen to launch the emoji selection list (some examples are shown in Figure 5-12).**

   This opens a number of emojis to choose from to express yourself in a more playful way.

FIGURE 5-12: Tap the emoji icon to send a playful smiley face or another emoji instead of (or in addition to) your words.

2. **Twist the Digital Crown button to select the right emoji that conveys your message or feelings, whether it's a smile, a silly tongue hanging out, a sad face, a heart, or something else.**

   The small green bar in the top right of the screen shows you where your list of options starts and ends. Leave the emoji on the screen for a moment to see how it'll animate once someone else receives it.

3. **When you find an emoji that fits the bill — maybe an animated thumbs-up or thumbs-down — tap Send in the top right of the screen to send it to the recipient.**

   If you decide against sending the emoji, tap Cancel in the top-left corner of the watch screen.

Although Apple removed the ability to send animated emojis from Apple Watch for some reason, at least there are multiple static emojis to send (Figure 5-13 shows you more examples).

FIGURE 5-13: Select a regular emoji (left) if you don't want to send one of the fancier animated ones (right).

Replying with Scribbles

If you don't feel like dictating your reply to a Message, you can use the Scribble feature to handwrite your words — letter by letter — using your fingertip.

To use the Scribble feature:

1. **Tap an incoming message.**

   A number of options appear.

2. **Tap the one that looks like a little finger swiping (left image of Figure 5-14).**

3. **Use your fingertip to write letters and numbers and the text version will appear above it (right image of Figure 5-14).**

   Tap the back button to go back one character or tap space for the spacebar. Tap Cancel in lower right of screen to cancel and start again.

4. **When you're done, tap Send in the top-right corner.**

FIGURE 5-14: Select Scribble to write your message. Using your fingertip, write each letter or number and you'll see the text version on top of the screen.

Responding with an audio clip

Unless you want to handwrite your response to a message using the Scribbles — which takes some time to do — you're likely to dictate a reply and hit Send. As covered previously in this chapter, if you're not in a place that prevents talking (like an important board meeting) you can review your words before you send a reply.

But wait there's another option! As I touch on in earlier sections, you can speak your reply to a message and send it as a voice clip. This is particularly useful for those times you want the recipient to hear your voice, like a kind of voicemail message.

Here's how to do it:

1. **Enable the voice recording feature.**

   As shown back in Figure 5-11, you enable this feature on your iPhone's Apple Watch app. Simply tap to open the app, go to Settings, and then Messages. Here you can select whether you want your watch to transcribe your voice into text only (Transcript), send everything as voice recordings (Audio), or enable prompts to give you either option after you reply to a message (Transcript or Audio).

2. **Read the message someone sent you by tapping on the green Messages app on your Apple Watch.**

3. **Tap the microphone icon and start talking.**

   When you're done, tap Done (easy, huh?).

   Now the screen will say "Send as Audio," or, if you opted for both, it will say "Send as Audio" or "Send as Text." Figure 5-15 shows what your message looks like before you send it, as well as what your recipient will see.

FIGURE 5-15: How a message appears as a voice recording or transcribed text (left and middle) and what your recipients see after they tap the message (right).

Sending a message

You can send a new message through Apple Watch in two ways: from the Messages app or by using Siri.

If you have an Apple Watch with cellular support, you don't need your iPhone nearby. If you don't, you will!

To send a message from your Apple Watch by using the Messages app, follow these steps:

1. **Press the Digital Crown button to go to the Home screen.**

2. **Tap the Messages app.**

   This launches the Messages app, where you can read previously sent or received messages.

3. **To send a new message, press and hold the screen until the words *New Message* pop up — as shown in Figure 5-16 — and then tap them.**

   Now you're ready to start a new message to someone.

4. **Tap Add Contact to select whom to send the message.**

5. **Tap Create Message, which lets you dictate your message, as also shown in Figure 5-16.**

   Again, you have an option for Apple Watch to transcribe your words into text, or you can send your message as a voice clip.

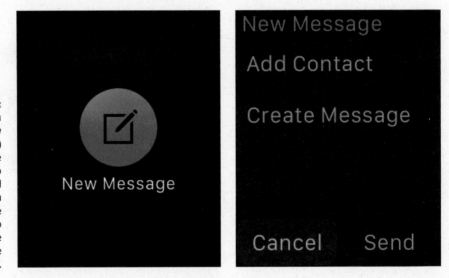

FIGURE 5-16:
Once you tap New Message (left) and choose a contact to whom to send a message, you can tap Create Message to compose the message (right).

To send a message on your Apple Watch using Siri, follow these steps:

1. **Lift your wrist and say "Hey, Siri" into your watch, followed by "Message ___ (person's name or number)."**

   You can also press and hold the Digital Crown button to activate Siri. Figure 5-17 shows a request example.

2. **The Messages app opens, and you should see the person to whom you want to send a message.**

   Dictate your message by tapping the microphone icon.

3. **Tap Send when you're done recording your message.**

   After a few moments, the person you're messaging with receive the text or audio clip.

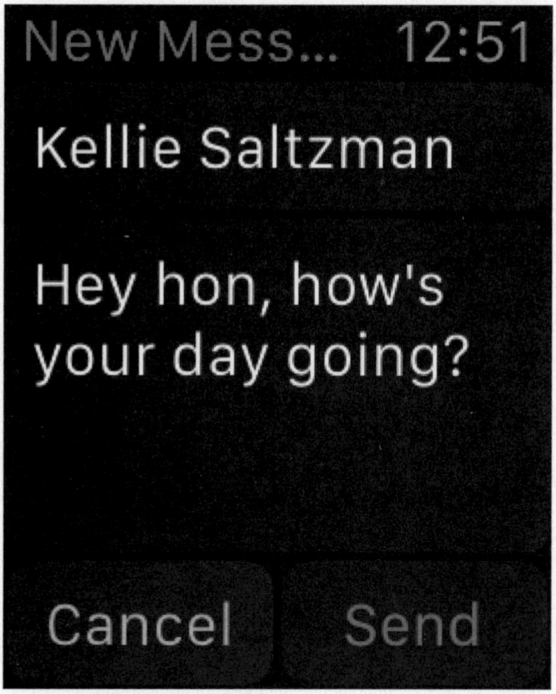

New Mess...   12:51

Kellie Saltzman

Hey hon, how's
your day going?

Cancel   Send

FIGURE 5-17:
Impress
your friends
by sending
a message
effortlessly —
all by using
your voice.

REMEMBER

A superfast way to send a message to someone through Apple Watch is to raise your wrist and say "Hey, Siri, message _____ (person's name)" and then say "_____ (message)." Siri shows you the message before you send it. See Chapter 7 for more ways to use Siri to help you perform tasks with your Apple Watch.

You can also start a message on Apple Watch and continue it on your iPhone. As you can with calls and emails, it's easy to transfer messages to your iPhone, where you can pick up right where you left off. Apple Watch is meant for quick interactions, not lengthy ones. See the "Handing Off a Call to Your iPhone or via Bluetooth" section to learn more about handing off calls to your iPhone from your Apple Watch.

Creating custom replies

If you find it convenient to select one of the many Smart Replies to someone through Apple Watch — like "Call you later" or "Thanks!" — keep in mind that you can create your own custom replies. You'll need your iPhone for this.

DIDYA KNOW?

Whenever you send a message over Apple Watch, it's mirrored with your iPhone, unless you want to turn that off. As you'll see in the figure, you can open your Messages app on iPhone and see the same conversation with someone as on your Apple Watch.

When synced, your Apple Watch and iPhone share the same messages, so you can start on your watch and continue on your phone, if you like. Gee, I hope Jacob likes my silly animated hamburger here (the eyes jiggle around).

To customize a reply:

1. **Open the Apple Watch app on your iPhone and tap the My Watch tab.**

2. **Choose Messages ⇨ Default Replies, and tap a default reply to change it.**

   Or tap Add reply at the bottom of the screen to create your own.

3. **To remove a default reply or change the order, tap Edit.**

   That's it!

Sharing your location from Apple Watch

Say you're chatting with a friend who you're supposed to meet up with but he or she can't seem to find you. Did you know Apple Watch lets you instantly share your location with someone? As shown in Figure 5-18, it's as easy as pressing down on your Apple Watch screen while in a messaging conversation with that person, and then tapping Send Location.

FIGURE 5-18: Instantly share your geographical location via Apple Maps by pressing down on the screen while chatting and selecting Send Location.

Your other options are Reply (send message back to them), Details (the Contacts info of the person you're chatting with), and Choose Language (to select from any preloaded languages).

Deleting a conversation from Apple Watch

It doesn't take up much room on your Apple Watch's storage, but if you want to remove a conversation after you've had it, it's quite simple, as shown in Figure 5-19.

1. **To delete a conversation, swipe left on the message.**

2. **Tap the red Trash bin icon.**

3. **Tap Trash again to confirm.**

FIGURE 5-19:
Delete
conversations
from Apple
Watch by
swiping to
the left on a
message and
tapping the
Trash can.

Sending Stickers, Taps, Kisses, Heartbeats, and More

Just like with your iPhone, Apple Watch lets you communicate via phone and text, but it can also do some things your smartphone can't do. Collectively, these actions fall under the Digital Touch features, available exclusively to Apple Watch.

Call it wrist-to-wrist communication.

Digital Touch allows Apple Watch wearers to connect with other Apple Watch wearers in fun, unique, and spontaneous ways. Specifically, Digital Touch offers six options, all shown in Figure 5-20.

>> **Sketch:** Draw something with your finger and the person you're sending it to sees it animate on his or her Apple Watch.

>> **Tap:** Send gentle (and even customizable) taps to someone to let that person know you're thinking about him or her.

>> **Kisses:** Send your sweetheart some virtual kisses directly to his or her Apple Watch by tapping the screen.

>> **Show anger:** Just like you can share your broken heart — wrist to wrist — you can also convey anger on your Apple Watch. Oooh, I know you want to try this one!

>> **Heartbreak:** Want to tell your lover you've been hurt? Your Apple Watch can do this for you with an animated broken heart.

>> **Heartbeat:** Your built-in heart rate monitor information is captured and sent to someone special so that person can feel it on his or her wrist.

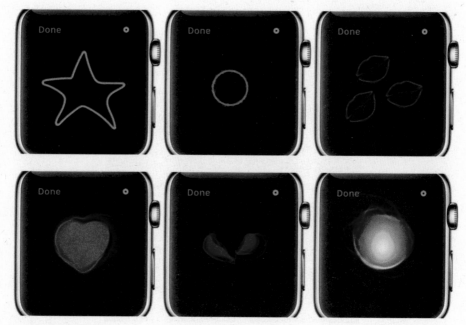

FIGURE 5-20: Use "Digital Touch" options to communicate from one Apple Watch to another. Clockwise starting at top left: Sketch, Tap, Kisses, Show anger, Heartbreak, and Heartbeat.

Sketch

You can draw on your Apple Watch screen and the share your creation — a sketch — with another Apple Watch owner. Apple Watch is little more than two square inches; therefore, don't expect to paint a masterpiece, but sketches are fun and creative ways to reach out to others. The person who receives a sketch can reply with one too, and you can chat back and forth like this to your heart's content.

To create and send a sketch from your Apple Watch, follow these steps:

1. Tap the Messages app and then select someone's name to reply to.

2. Select the icon that looks like a heart, with two fingers inside.

This is for the half-dozen Digital Touch options.

3. **Sketch something with your fingertip and when you lift up, it will send the sketch to the other person.**

   You can draw a smiley face, a star, a heart, a flower, a sun, a fish, written-out words, or anything else you can think of. See Figure 5-21 for an example. Because your friend is wearing an Apple Watch, that person sees the drawing appear on his or her wrist just as you drew it. He or she knows it's from you because your name is in the top-left corner (which is the same for a tap and a heartbeat).

FIGURE 5-21: This is a (poorly drawn!) example of what you can sketch and send to a friend's Apple Watch.

When sending a sketch, you can tap the small circle at the top right of the Apple Watch screen to change colors. See Figure 5-22 for a look at the seven colors offered in the palette.

Tap

Another unique way to use Digital Touch on Apple Watch is to send a tap to someone. As with the other Digital Touch features, that person needs to have an Apple Watch too. A tap is similar to a tactile version of a Facebook "poke" — a kind of "Hey, I've been thinking about you" type of notification.

To send a tap from your Apple Watch, follow these steps:

1. **Tap the Messages app and then select someone's name to reply to.**

2. **Select the icon that looks like a heart, with two fingers inside.**

   This is for the half-dozen Digital Touch options.

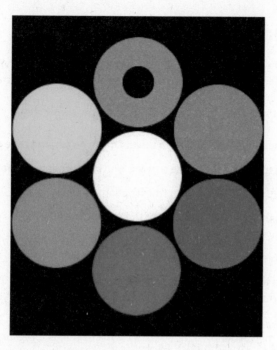

FIGURE 5-22:
You're not limited to just one color for a sketch. Tap the small circle, select another color, and then draw away.

3. **Tap the screen one or more times, such as two quick taps in the top-left corner and a single press on the bottom right of the screen.**

Tap the small circle in the top right to change the color. Stop tapping to send.

After you complete those taps, the person you're tapping will see and feel the tap on his or her Apple Watch — visually presented as small and large rings that disappear (as shown in Figure 5-23) — and he or she knows it's from you because your name is in the top-left corner (which is the same for a sketch and a heartbeat).

The recipient can "play" the pattern again by tapping in the upper-right corner. If and when the person you're tapping with replies with a Digital Touch, he or she can choose to reply with something else, such as a sketch or a heartbeat instead of another tap.

FIGURE 5-23:
Send a custom tap to someone else that has an Apple Watch and that person should see and feel it on his or her wrist.

Heartbeat

Sending your heartbeat is another way to reach out and flirt with someone — from your Apple Watch to another person's Apple Watch. Is that romantic or what!

To send a heartbeat from your Apple Watch, follow these steps:

1. **Tap the Messages app and then select someone's name to reply to.**

2. **Select the icon that looks like a heart, with two fingers inside.**

   This is for the half-dozen Digital Touch options.

3. **Press two fingers on the screen at the same time.**

   Apple Watch's heart rate monitor (which is underneath the watch) immediately starts calculating your heart rate.

   After a few seconds, your heartbeat — as shown in Figure 5-24 — is sent to (and felt by) your significant other or friend. He or she knows it's from you because your name is in the top-left corner (which is the same for a sketch and a tap).

   Okay, this might sound a little gimmicky, but it might just brighten up your friend's or loved one's day if he or she is having a rotten one. Just don't send it to the wrong person, such as your boss or your best friend's spouse — in case someone gets the wrong idea.

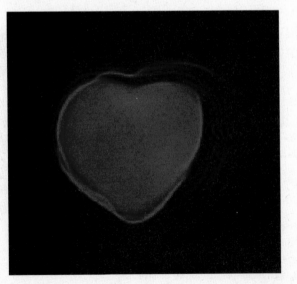

FIGURE 5-24: Send your heartbeat to a friend's or your better half's Apple Watch, which they'll see as a beating heart. You just need two fingers to do so.

A few more Digital Touch options

Didn't see what you wanted to send your friends and loved ones in the previous sections? Try the following options:

>> **Send a Kiss:** Similar to Heartbeat, tell someone you're thinking of them by sending kisses. Follow the instructions in "Sending a message" to select a person to message with, tap the Digital Touch icon, and to send kisses, tap two fingers on the screen one or more times. Stop tapping to send.

>> **Break a heart:** Time to channel your inner Taylor Swift and tell someone they've broken your heart (yes, I went there). Follow the instructions in the "Sending a message" section to select a person to message with, tap the Digital Touch icon, and then place two fingers on the screen until you see and feel your heartbeat. Drag down to send.

>> **Show anger:** Don't know how to communicate your fury? Apple Watch to the rescue. Follow the instructions "Sending a message" to select a person to message with, tap the Digital Touch icon, and then touch and hold one finger on the display until you see a flame. Lift to send.

Stickers (including Animojis, Memojis), handwritten messages, and Tapback

As you can tell, Apple went all-out with its Messages app for Apple Watch. There's even more you can do, if you can believe it.Here are a couple of extra goodies to play around with.

Stickers

Want to send someone a fun sticker to make their day? It's quick and easy.

1. **Tap someone's name in your Messages app and hit Reply.**

2. **Tap the smiley icon, which has emojis inside.**

3. **Scroll to the bottom by swiping down or twisting Digital Crown, and select the word *Stickers*.**

   Recent stickers and handwritten messages sent from your other devices appear here (swipe sideways to see your Recent ones) so you can send them again (or to someone else) from Apple Watch. Figure 5-25 shows an example.

FIGURE 5-25:
Send fun
stickers or
handwritten
messages from
other Apple
devices right
from Apple
Watch.

Animojis and Memojis stickers

As you likely know from your iPhone, Animojis — short for *animated emojis* — uses the iPhone's front-facing camera to create a cartoon emoji that mimics your facial expressions in real time. Memojis, on the other hand, are customizable Animojis, after you've tweaked your head shape, facial features, hair, and more.

Adding Animojis/Memojis to messages on Apple Watch is as easy as adding a regular sticker, but you need to make sure your Apple Watch is running watchOS 6 or later (Settings ⇨ General ⇨ Software Update). When you scroll down the list of available stickers — by swiping down the screen or turning the Digital Crown — you see the Animojis (the left side of Figure 5-26) and Memojis (if you've created any) just below the regular stickers (the right side of Figure 5-26). Tap which one you want to send. That's it!

Tapback

You can respond to someone's message without typing out a full response. Think of tapbacks as messaging shortcuts. To quickly send one:

1. **Open the Messages app on Apple Watch.**

2. **Double-tap a message.**

3. **Choose a Tapback, like thumbs-up or a heart (see Figure 5-27).**

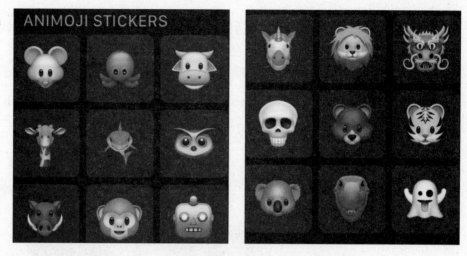

FIGURE 5-26: Have fun exchanging Animojis (pictured here) or custom-made Memojis inside of Apple Watch's Messages app,

FIGURE 5-27: Thumbs up? Thumbs down? You can choose from many Tapback options.

Enabling and Using the Walkie-Talkie Feature

 Walkie-Talkie is a fun way to communicate between Apple Watch wearers. As you might expect, it lets you quickly chat with someone, wrist to wrist, using your voice (see Figure 5-28).

FIGURE 5-28: 10-4, good buddy? You can use the Walkie-Talkie feature to chat with someone via their Apple Watch.

REMEMBER

The Walkie-Talkie app isn't available in all countries or regions. But no, you don't need to be near each other like old-school walkie-talkies (like in the Stranger Things TV show)!

To get going, you have to fulfill two requirements:

» You and your friend both need any version of Apple Watch, as long as you have the watchOS 5 operating system or later.

» You also both need to set up the FaceTime app on your iPhone and be able to make and receive FaceTime audio calls. See the earlier section in this chapter "Outgoing calls with FaceTime audio."

Adding friends to Walkie-Talkie

To do so:

1. **Open the Walkie-Talkie app on your Apple Watch (it's yellow and black).**

2. **Press the yellow "+" sign and choose a contact.**

   Wait for your friend to accept the invitation. The contact card will stay gray and is labeled "invited" until your friend accepts.

3. **After he or she accepts, the contact's card turns yellow and you and your friend can now talk instantly.**

   See Figure 5-29 for what this looks like.

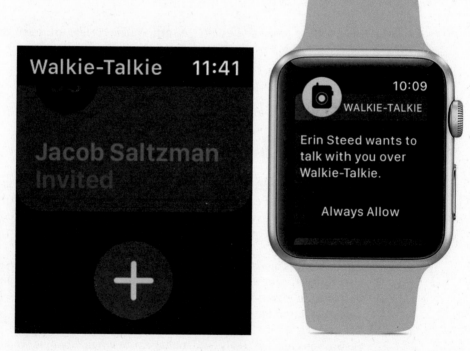

FIGURE 5-29: Send an invite to someone over Walkie-Talkie (left) and your friend's wrist will display your request (right).

To remove a friend, open the Walkie-Talkie app, swipe left on the friend, then tap the red X. Alternatively, you can open the Apple Watch app on your iPhone, tap Walkie-Talkie ⇨ Edit, tap the red – (minus), and then tap Remove to confirm.

Starting a Walkie-Talkie conversation

Now for the fun stuff:

1. **Open the Walkie-Talkie app on your Apple Watch.**

2. **Tap a friend's contact.**

3. **Touch and hold the talk button, and then say something.**

   You might see the word *connecting* on the screen as the watches are attempting to wirelessly connect. After the connection is made, your friend can hear your voice and talk with you instantly.

4. **To talk over Walkie-Talkie, touch and hold the talk button, then say something; when you're done, let go.**

   Your friend instantly hears what you said. To change the volume, turn the Digital Crown.

Need a little peace and quiet? To turn Walkie-Talkie off, open the app and turn Available off or on.

TIP

Sending, Receiving, and Managing Emails on Apple Watch

Apple Watch would be a half-baked product if you could only use it for reading messages and not email. Thankfully, it delivers a decent mail experience right on your wrist. You can read and manage your mail, reply, or even compose a response to an email.

Apple Watch's inbox is synced with your iPhone, so you can browse by date, sender name, titles, contents, and attachment/VIP status or by the default All Inboxes, as shown in Figure 5-30. This is especially important for those who rely on email communication for work.

Before you can read, reply, and manage your email on Apple Watch, you'll want to ensure it's synced with the mail account(s) on your iPhone. As shown in Figure 5-31, you should open the Apple Watch app on iPhone, tap Mail, and ensure your accounts are listed. Here you can also tell your iPhone how you'd like to be notified of new emails on your Apple Watch (tap Custom).

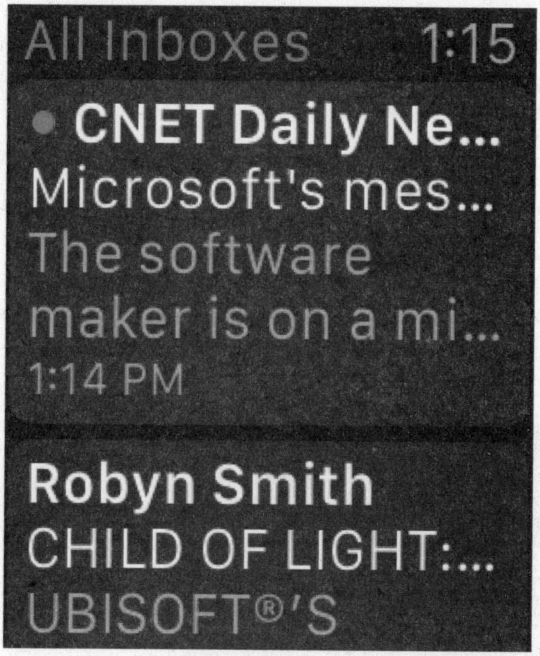

FIGURE 5-30:
You can manage your inbox from your wrist.

FIGURE 5-31:
The Mirror My iPhone option on the Apple Watch app.

Reading and acting on an email

To read and act on an email message, follow these steps:

1. **Press the Digital Crown button to go to the Home screen.**

2. **Tap the Mail app.**

    Or raise your wrist and say "Hey, Siri, Mail." Either action launches the Mail app and takes you right to your inbox.

3. **Use the Digital Crown button or your fingertip to scroll up and down through your emails — as shown in Figure 5-32 — and then tap the subject line to open one.**

   The email you selected fills up your Apple Watch screen, hiding the others.

4. **Once in an email, scroll down to the bottom of the message and tap Reply, as shown in Figure 5-33.**

   You can also turn the Digital Crown to scroll to the bottom. You can also press and hold Reply to see more options: Reply, Flag, Unread (mark as unread), or Archive (save for later).

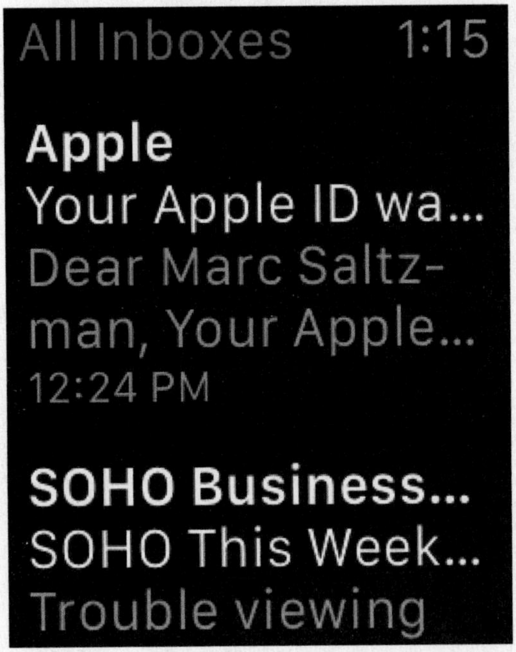

All Inboxes   1:15

Apple
Your Apple ID wa...
Dear Marc Saltz-
man, Your Apple...
12:24 PM

SOHO Business...
SOHO This Week...
Trouble viewing

FIGURE 5-32:
Navigating
your inbox
using the
Digital Crown
button.

Replying to an email

If you want to reply to an email, you've got a few choices, as shown in Figure 5-34:

» **Microphone:** To dictate your email reply, tap this option. It will transcribe your voice into text. Tap Done. If you see that the words onscreen look good, tap Send. If not, hit Cancel.

» **Scribble:** Reply using the Scribble feature, which means you can use your finger to write a reply. See the section "Replying with Scribbles" for full instructions.

FIGURE 5-33:
Swipe down
in an email
message to get
to the bottom
and then tap
Reply (left).
Pressing and
holding Reply
pulls up these
additional
options (right).

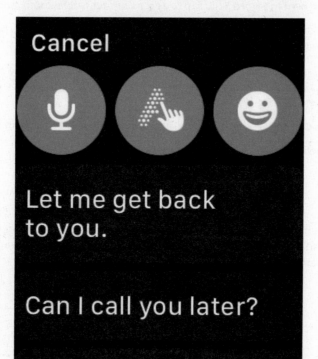

FIGURE 5-34:
Tap Reply at
the end of an
email message
and you've
got some
options here:
microphone
(voice-to-text),
scribble, emo-
jis, or default
replies.

>> **Emojis:** Select the smiley face to open the many emoji examples for you to send.

>> **Default replies:** Tap one of the preset phrases for a quick reply. You can edit these on your iPhone, in the Apple Watch ⇨ MyWatch ⇨ Mail area.

Composing an email

Although this feature wasn't available when the Apple Watch first debuted a few years ago, you can now create an email from scratch on Apple Watch! And you might not know it's even available because there isn't a "Compose Email" option, nor does Apple cover it on the web pages devoted to Apple Watch.

No worries: I've got your back! Here's how to compose an email on Apple Watch:

1. **Open the Mail app.**

   You will likely see emails in your inbox.

2. **Press and firmly hold the screen and you'll feel a slight vibration.**

3. **Tap** *New Message*.

4. **As shown in Figure 5-35, tap to add a Contact, Subject line, and Message.**

   - For Contacts, choose the recipient from your Contacts, or verbally say a name as shown in the left image of Figure 5-36.

   - Tap Subject, and you'll have the option to dictate, Scribble, use an emoji, or choose from a preset/default list of Subject lines as shown on the right image of Figure 5-36.

5. **Tap Create Message and you'll see similar options to those in Step 4.**

6. **When you like what you see, press Send.**

That's it!

Oh, wait, you say. What about Siri? Can't you raise your wrist and ask Siri to compose an email right from Apple Watch? Regrettably, no. If you try that, you'll hear Siri say she can help you compose an email on your iPhone, if you like. But not on Apple Watch — or perhaps, not yet.

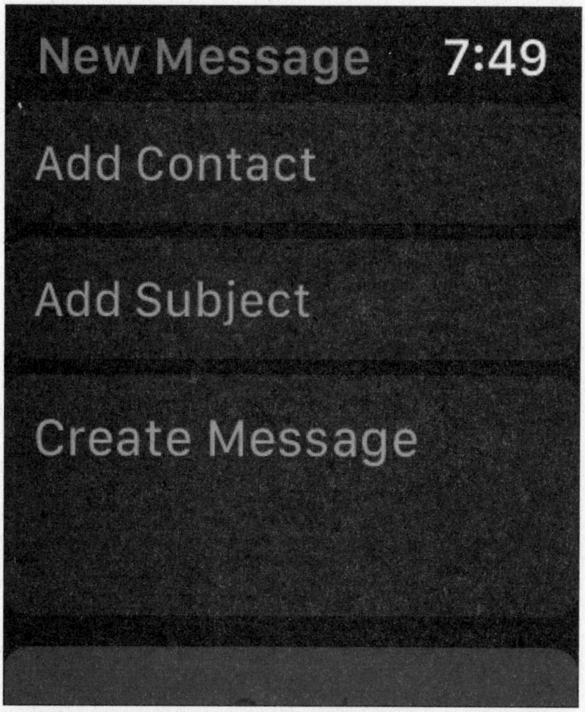

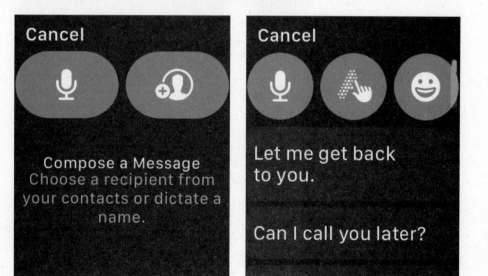

New Message   7:49

Add Contact

Add Subject

Create Message

FIGURE 5-35: Composing an email on Apple Watch is as easy as pressing and holding the screen in the Mail app. Tap New Message and then fill out the rest.

Cancel

Compose a Message
Choose a recipient from your contacts or dictate a name.

Cancel

Let me get back to you.

Can I call you later?

FIGURE 5-36: Options to send an email include transcribing/ choosing a contact (left). Options for the Subject line and Message body are just like messaging options (right). Cool, huh?

» Accessing weather, UVI and AQI index, and wind speed

» Following stock info and more

» Adding information as complications

» Using Dock to quickly open your favorite apps or go from one app to another

» Viewing notifications on Apple Watch

» Mastering the calendar for your wrist

» Tapping the Calculator app

» Scheduling and seeing Reminders

» Setting and listening to Voice Memos

» Navigating the Maps app, including the new Compass feature (Apple Watch Series 5 only)

Chapter **6**

In the Know: Staying Informed with Apple Watch

When people ask me, "Why do I need a smartwatch?" I usually answer with something like "convenience." A wearable device on your wrist is ideal for quick bits of information, when and where you need it. Rather than pulling out your smartphone or your tablet, you can simply glance down with your eyes to get what you need while on the go.

This chapter focuses mostly on how Apple Watch shows you information that's relevant to you. For example, you can access weather information for today or for the next week, or even for another city. Or multiple cities. Do you like playing the market or want to see how your own public company is doing? Select which company's stock price and performance are important in your world — all updated in real time.

Then I describe how to add this information to your watch face as one of the complications options. I also cover how to access notifications on Apple Watch for a number of installed apps. I round it all off with using Apple Watch for calendar alerts and accessing maps on your wrist.

Accessing Real-Time Weather, UV Index (UVI), and Wind Speed

Apple Watch comes with many built-in Apple Watch apps (see Table 1-1 in Chapter 1 for a list of them), and Weather is one of them.

Sure, you can choose not to see this information on your smartwatch if you're not interested — simply unselect it in the Apple Watch app on your iPhone, as covered in Chapter 11 — but this information should be timely and interesting for many (hence, Apple adding them in without your having to download specific apps).

Apple also added UV (ultraviolet) index info, wind speed, and in some cities, the Air Quality Index (AQI), too.

The Weather app displays your current location's temperature and a graphical representation of expected precipitation and conditions over much of the day (by the hour) as well as a week-ahead view — and for multiple locations if desired.

To use the Weather app on your Apple Watch, follow these steps:

1. **Press the Digital Crown button to go to the Home screen.**

2. **Tap the Weather app.**

   Once the app launches, you should see the temperature and weather for your current location, as shown in Figure 6-1. The temperature is in the center of the app, surrounded by a 12-hour look at the day and an associated icon, such as a sun, a cloud, a raincloud, snowflakes, and so on. If you live in the United States, the temperature is displayed in Fahrenheit (a choice you make when setting up the watch); it's displayed in Celsius in Canada and other countries that use the metric system or if you make that choice yourself. The name of the city appears in the top left of the screen and the current time in the top right.

FIGURE 6-1:
Whether it's
a local city or
faraway cities,
get real-time
weather
conditions
from the
Weather app.

3. **Swipe down with your fingertip or twist the Digital Crown button to scroll down the app.**

   This shows you a ten-day forecast, as shown in Figure 6-2.

   The further down you swipe or twist the Digital Crown button, the further ahead in the week you go. You can see each day's high and low temperatures and predicted precipitation — all courtesy of the Weather Channel. Scroll back up to the top of the Weather app by twisting the Digital Crown or use your fingertip to swipe up.

4. **Press the screen (Digital Touch) for more info, such as chance of rain (percentage).**

5. **To see other cities you've selected to track, swipe to the right or left to pull up the weather there, as shown in Figure 6-3.**

   You can change the cities you follow in the Apple Watch app on iPhone; see Chapter 11 for more on doing this.

   Small white dots at the bottom of the app's screen show you how many pages (cities) you've selected. You can add or remove cities in the Apple Watch app on the iPhone. Open the app and select My Watch, followed by Settings and then Weather. Figure 6-4 shows examples of weather conditions you can access.

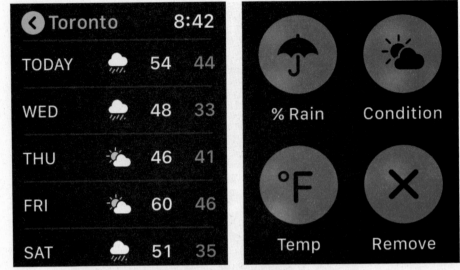

FIGURE 6-2: The more you scroll down, the more weather information (left) you see. Or press the screen firmly and select additional information (right).

FIGURE 6-3: Flick to the left or right to access weather information in other cities. Customize what you see by using the Apple Watch app on iPhone.

FIGURE 6-4:
The watchOS 5 operating system and later has UV index, wind speed, and, in some supported cities, air quality info.

REMEMBER

Don't forget you can use Siri to call up weather information. Simply raise the watch to your mouth, say "Hey, Siri" and or press and hold the Digital Crown button, followed by a question, such as "What's it like outside?" or "Do I need a raincoat?" Or give a command, such as "Tell me the weather." Actually, you can just say "Weather." If you don't specify a location — like the examples provided here — Siri assumes you want to know your local information. See Chapter 7 for more on using Siri to help you perform tasks with your Apple Watch.

Following Stock Information and Much More

To keep an eye on the stock market, Apple Watch offers you a quick glance at any public company's stock price and performance for when (and where) you'd like it. It's similar to the Stocks app on iPhone but tailored to the smaller Apple Watch screen. The process is similar to looking at the weather and just as customizable.

To use the Stocks apps on your Apple Watch, follow these steps:

1. **Press the Digital Crown button to go to the Home screen.**

2. **Tap the Stocks app.**

If you didn't tweak what stocks you'd like to see by going into the Stocks area of the Apple Watch app on iPhone, see Chapter 11 for more on doing this. Until you customize what stocks you want to keep an eye on, you will see the current value of major stock indices, such as the Dow Jones Industrial Average, as well as such companies as Apple and Starbucks, as shown in Figure 6-5.

Each company you follow on the stock market is listed by its traded name (such as AAPL for Apple), the current stock price underneath the name (such as $222.15), and whether the stock is up (in green) or down (in red) and by how much, such as a green +2.20%.

FIGURE 6-5:
See all kinds of
stock price and
performance
information
on any
publicly traded
company.

3. **You can learn more information about each company or exchange by tapping its name (such as DOW J).**

   Figure 6-6 shows an example.

FIGURE 6-6:
Despite its
small screen,
Apple Watch
can give
you a ton of
information
about the stock
market, such as
this snapshot
of the Apple
stock over the
past day.

This includes the full company name, point and percentage changes, and market cap.

You can have a number of companies listed, all organized within the Apple Watch app on iPhone. See Chapter 11 for more on third-party apps.

The Stocks app on iPhone will give you an even deeper dive, such as the indices or company's performance over the past day, week, month, or six months (see Figure 6-7).

4. **Swipe to the right at any time to go back to the main Stocks app screen with multiple indices and companies listed.**

You can also ask Siri to tell you the stock price of a given company or the performance of a stock index. You can access Siri by saying "Hey, Siri," followed by your question or command, or press the Digital Crown button in any app you're in to ask Siri about a particular company.

FIGURE 6-7: View a snapshot of company stock or index performance over the past hours (left). Or open the Stocks app on iPhone for an in-depth view of the stock index or company, courtesy of Yahoo! (right).

Adding Weather and Stocks to Your Watch Face Screen

Apple Watch lets you customize your watch face in a variety of different ways — see Chapter 4 for more on doing that — including the option to add complications.

To refresh your memory, *complications* are extra information you want visible on the watch face itself — such as weather and stocks — so you can see it when you're checking the time. Depending on which watch face you choose, complications are usually reserved for the four corners of the screen and perhaps the bottom center. Tapping the piece of information provided, such as the weather, opens the corresponding app for a deeper look.

To add weather and stock information — or other complications, such as moon phase, sunrise and sunset information, alarm clocks, timers, and so on — to your watch face, follow these steps:

1. **Press and hold the screen when viewing a watch face.**

   This enables Force Touch and launches the Faces gallery.

2. **Swipe left or right to select a clock face you like.**

3. **Tap Customize near the bottom of the screen to add more information to each watch screen.**

   You can adjust the color of the hands and add or remove details, such as the second hand, by twisting the Digital Crown button.

4. **Swipe to the left to go to the last customization screen, which takes you to complications, as shown in Figure 6-8.**

   This is where you can add weather and stock quotes to the watch face you've selected.

5. **Tap the areas you want to customize (shown in green in Figure 6-8) and then twist the Digital Crown button to select what you'd like.**

   Repeat the process by tapping the other areas of the face you want to change. Remember, you won't have the same options for all watch faces. The watch face in Figure 6-8 doesn't allow for stock information, for example, but other watch faces do.

6. **Press the Digital Crown button when you're done customizing.**

7. **After you make your choices, press the Digital Crown button again to go back to the Home screen.**

   After you've set your watch face, don't forget you can tap on each of the complications, such as weather or stock quotes, to go directly to the relevant app. See Chapter 4 for more on complications.

You can take a screenshot of whatever you're doing on Apple Watch. Simply press the Digital Crown button and the side button at the same time and you should see the screen briefly flash white. Now check your iPhone's Photos app. Your newly captured image should be there.

TIP

Using Dock on Apple Watch

Covered briefly in Chapter 3, Dock lets you quickly open and cycle through your most recently used apps (or your favorite apps) when you're on the go.

In other words, it's a super convenient shortcut to what matters most to you.

Launching Dock and more

Here's how to launch Dock, screen between apps, launch one, and change what you see there.

1. **Press the side button.**

   This activates Dock on Apple Watch.

2. **Swipe up or down with your fingertip or turn the Digital Crown.**

    This cycles through the last apps you opened or your favorite apps (see the next section on how to customize this). Figure 6-9 shows an example of what it looks like to cycle through some apps.

3. **To close an open app in from Dock, swipe to the right and press the big red "X."**

4. **Tap the name of the app to open it full-screen.**

    If you scroll all the way down to the bottom of the screen, you can tap All Apps to go to the Home screen.

5. **To close Dock, press the side button again.**

FIGURE 6-9:
Here's a look at Dock on Apple Watch, displaying recently used apps or your pinned favorites (up to ten).

Customizing Dock

To choose which apps appear in Dock — up to ten of your favorites — follow these steps:

1. **Grab your iPhone and open the Apple Watch app.**

2. **Tap My Watch, then Dock.**

    Here you can choose your favorite apps.

3. **Tap Edit and then add or remove apps (see Figure 6-10).**

    To remove apps, tap the red –, then tap Remove. To add apps, tap the green +.

4. **To rearrange apps, touch and hold next to an app, then drag up or down.**

5. **Tap Done.**

    This saves your changes.

Make sure that Favorites is selected to have them appear on Dock.

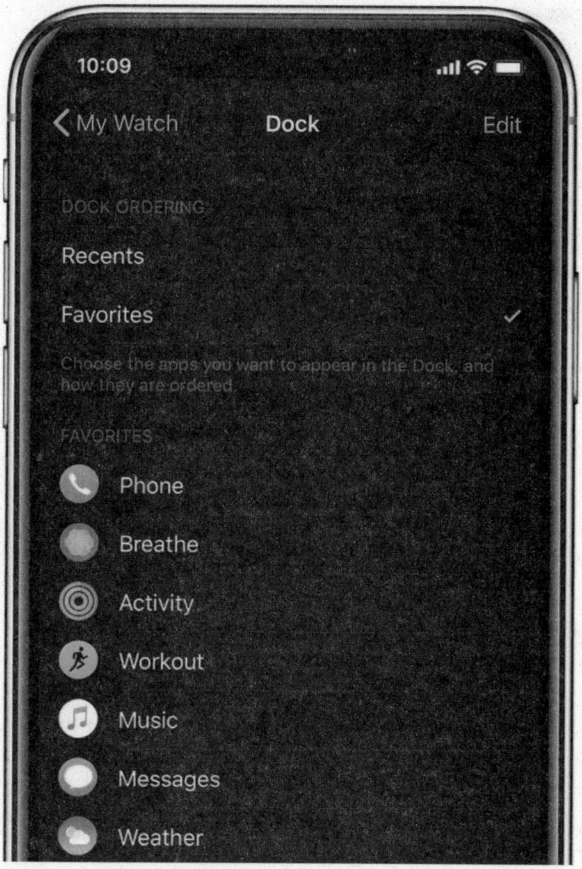

FIGURE 6-10:
On the iPhone Apple Watch app, you can select what apps to see when you activate Dock.

Mastering Notifications on Apple Watch

Just as many preinstalled Apple Watch apps support notifications (which are on by default), many third-party apps have them too — depending on whether app developers provided them. This section shows how to navigate notifications to meet your situation.

How notifications work

Notifications pull information from supported apps — for example, if someone liked your photo on Instagram, if your favorite game wants you to know someone is attacking your virtual kingdom, or if the New York Yankees won the doubleheader. Likewise, a news outlet, such as USA TODAY, might let you know the president is about to give a speech or if the weather is taking a turn for the worse in your area. Figure 6-11 gives an example of a notification tied to the USA TODAY app.

FIGURE 6-11:
USA TODAY,
and many
other
third-party
news apps,
push headlines
and other
notifications
to your wrist
so you can
glance at your
Apple Watch
to see timely
info that's
meaningful
to you.

USA TODAY 1:12

Kim Jong Un cancels Russia trip

It would have been the reclusive ruler's first official foreign visit.

But first a quick caveat: Notifications can go to either your Apple Watch or iPhone, but not both. If your iPhone is unlocked, you'll get notifications on your iPhone rather than your Apple Watch. But if your iPhone is locked, asleep, or off altogether, you'll get notifications on your Apple Watch (unless your Apple Watch is locked with your passcode). Figure 6-12 illustrates this.

If you have an Apple Watch without cellular connectivity, notifications are pushed to your watch from a nearby iPhone or over Wi-Fi. Otherwise, if you own a GPS + Cellular model and pay for the service, you can get notifications to your wrist directly from apps, over the Internet.

REMEMBER

Chapter 11 explains how to enable or disable individual app notifications in the Settings area of the Apple Watch app on your iPhone.

Accessing the notification settings

Quite simply, you can use the Apple Watch app on your iPhone to access notifications and choose which apps give you notifications. You can also select what information you'd like to be notified about per app. So for example, the right image of Figure 6-13 shows how to customize the Calendar app, which you simply do by tapping Custom to choose when to be notified. The Notification Center also includes apps from third-parties.

FIGURE 6-12:
In the Apple
Watch app
on iPhone,
you can select
where you see
your notifica-
tions, per app,
and whether
you want them
on your wrist
or only on your
iPhone.

FIGURE 6-13:
In the Apple
Watch app
on iPhone,
select if you'd
like notifica-
tions on your
Apple Watch
(left). You can
choose which
apps you'll
be notified
about (middle).
You can also
customize
notification
information, in
this instance
the Calendar
app (right).

Now, you should receive a gentle *tap* on your wrist to tell you about the news based on the apps you choose to give you notifications.

TIP

If you're not feeling the notifications on your wrist, you can dial up extra vibration by selecting Prominent Haptic within the Apple Watch's Sound and Haptics area in the Settings app. Now those buzzes on your wrist will feel more pronounced.

Viewing notifications

Just like on your iPhone or iPad, you swipe down from the top of the Apple Watch Home screen to access your Notification Center. Here, you can scroll up or down with your finger, or you can twist the Digital Crown button to see your next calendar appointment, how your stocks are doing, and perhaps a look at traffic on the way to the office.

To view notifications on your Apple Watch, follow these steps:

1. From the watch face screen, press at or swipe down from the top of the screen to open the Notification Center.

Remember to start swiping down from the very top of the watch case — on the rim — to successfully pull up the information you want. Swiping from the middle of the screen doesn't work. Or if you press down too close to the center of the watch face, you'll pull up all your watch faces rather than notifications. It may take a bit of trial and error until you find the "sweet spot."

You first see a summary of the notification, depending on the app. You may need to keep your arm raised to see more details on the notification.

2. Use your fingertip to swipe up and down for additional information, or twist the Digital Crown button.

To go back to your Home screen, press the Digital Crown button.

3. View or dismiss notifications (to dismiss, swipe up).

When you receive a notification, such as a calendar appointment, a red dot appears on your watch face. You first see a quick summary of the notification, followed by more info (depending on the app and notification). When you dismiss notifications by swiping up, they're also dismissed from your iPhone. You can also delete a notification by swiping to the left and tapping the large X.

Changing how you get notifications on your Apple Watch

Apple Watch isn't a one-size-fits-all scenario. Different wearers will want to be notified about different apps, and in different ways. Concerning the different ways, here's how to make some changes.

1. **Touch and hold the top of the watch face, or swipe down from the top of the watch face to open the Notification Center.**

2. **After the Notification Center appears, swipe down.**

3. **Swipe left on a notification, tap the three little dots, then select an option (see Figure 6-14):**

   - *Deliver Quietly:* If you don't want to hear sounds or feel haptic alerts for that app, tap Deliver Quietly. Notifications will now go directly to Notification Center on both your Apple Watch and iPhone, rather than produce a sound or haptic alert.

   - *Turn Off on Apple Watch:* If you don't want to get notifications for an app, tap this option.

4. **If you change your mind about an app, swipe left on a notification from the app, tap the three dots, and then choose Deliver Prominently.**

FIGURE 6-14:
Choose how to receive app notifications right from Apple Watch.

TIP

Want more privacy? When you raise your wrist to view a notification, you see a quick summary, and then full details a few seconds later. To prevent the full details from appearing in the notification, open the Apple Watch app on your iPhone, then tap the My Watch tab. Tap Notifications, then turn on Notification Privacy. Now, when you receive a notification, you need to tap it to see the full details.

Accessing Your Calendar on Apple Watch

Because your Apple Watch is connected to the Internet via cellular or Wi-Fi — or at least wirelessly tethered to your nearby iPhone through Bluetooth technology — you can access handy calendar information on your wrist. In fact, the Calendar app not only syncs with your iPhone, but it also syncs with iCloud if you use Apple's popular cloud service to store and access information. This section covers how to navigate your calendar, how to respond to a calendar or appointment request, and how to respond to an appointment request through notifications.

Navigating the Calendar app

Featuring day, week, and month views — including support for reminders, invitations, and notifications — the Calendar app on Apple Watch shows you a list of upcoming events.

To use the Calendar app on your Apple Watch, follow these steps:

1. **Press the Digital Crown button to go to the Home screen.**

2. **Tap the Calendar app.**

   This launches the Calendar app. By default, you should see the current Today view, with your upcoming events listed in chronological order.

3. **Use your fingertip to scroll down to see future dates or twist the Digital Crown button toward you.**

   Figure 6-15 shows an example of an upcoming event.

   This experience is similar to the Calendar app on iPhone. The current time is also listed in the top right of the Calendar app.

4. **In the main Calendar app screen, press and hold the screen to access the Month view; if you want to look at a date (such as Friday, October 16), just tap the day on the calendar for a Day view.**

   See Figure 6-16 for a look at the Month view. In this view, you can swipe left or right to move forward or backward through time or twist the Digital Crown button if you prefer.

TIP

Just like you can set Reminders on your iPhone, iPad, or iPod touch, you can activate Siri on your Apple Watch and say something like, "At 6 p.m., remind me to call Mom." The watch doesn't make a calendar entry for this event, but you're reminded with a sound, a vibration on your wrist, and a text. Remember, Apple Watch has no keyboard, so you must dictate the reminder. See Chapter 7 for more ways to use Siri to help you accomplish tasks with Apple Watch.

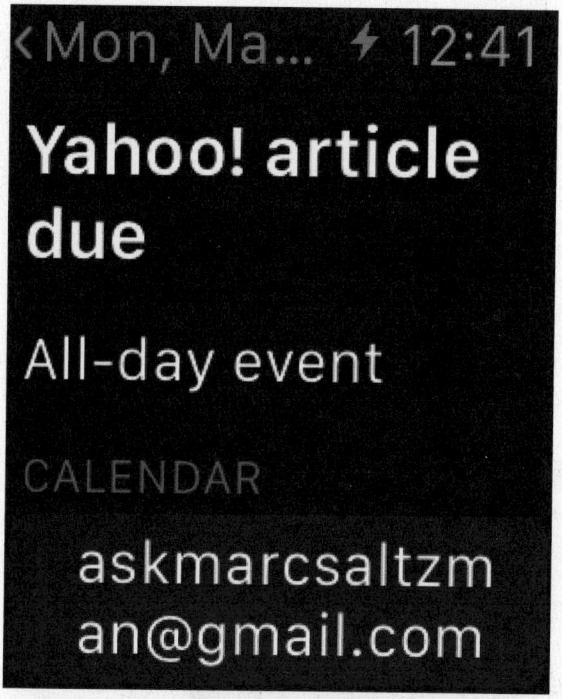

FIGURE 6-15:
The Calendar app offers a look ahead — beyond the current date — if you want to preview your upcoming week.

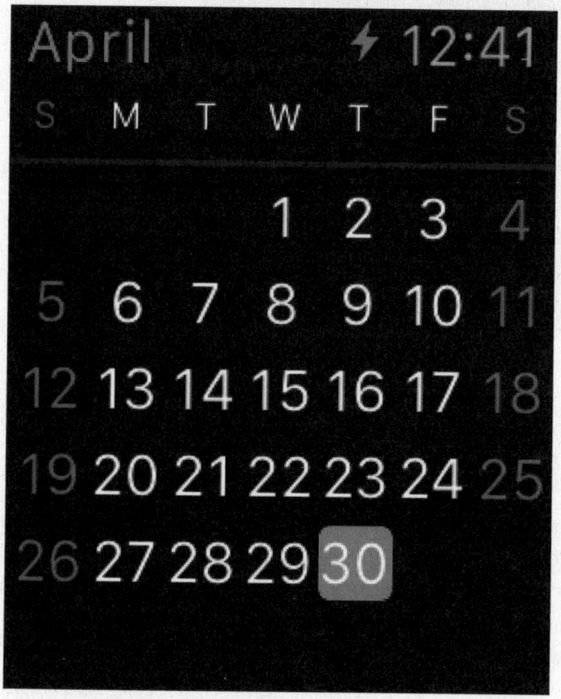

‹Mon, Ma... ⚡ **12:41**

Yahoo! article due

All-day event

CALENDAR

askmarcsaltzm an@gmail.com

FIGURE 6-16:
See what's happening a month at a time and then tap a day for a chronological view of what events you have on a given day.

April ⚡ **12:41**

S	M	T	W	T	F	S
			1	2	3	4
5	6	7	8	9	10	11
12	13	14	15	16	17	18
19	20	21	22	23	24	25
26	27	28	29	30		

5. **Press firmly on the screen to access various view options (see Figure 6-17):**

   - *List view:* Displays your appointments in a list.

   - *Today view:* Shows what's happening today if you're already looking at Month view.

   - *Up Next:* Shows details on your next appointment.

6. **Create a calendar appointment on your Apple Watch.**

   To do this, you'll need to use Siri. Simply ask your personal assistant to make a calendar entry on a given date and time.

FIGURE 6-17:
When in the Calendar app, press firmly on the Apple Watch screen to pull up extra view options.

Responding to a calendar/appointment request

Say you received an invite or you're reviewing one; Apple Watch makes it easy to reply without needing your iPhone. Plus you can perform other tasks, such as sending an email quickly or sharing map info.

1. **Tap a calendar invite, as shown in the top left image of Figure 6-18.**

   You see the date and time of the appointment, and which calendar/email account it's tied to (in case you have different accounts).

2. **Press and hold the screen.**

   Now you can see more info about the person who invited you, and scroll down to Accept, Decline, or choose Maybe as shown in the top right and bottom left images of Figure 6-18.

3. **Press and hold on the name of the organizer.**

   Other options (the bottom right image of Figure 6-18) appear, including sending the invitee an email or getting directions to where you're going, which also estimates driving time from your current location, via Apple Maps (see the section "Navigating the Maps App" later in the chapter).

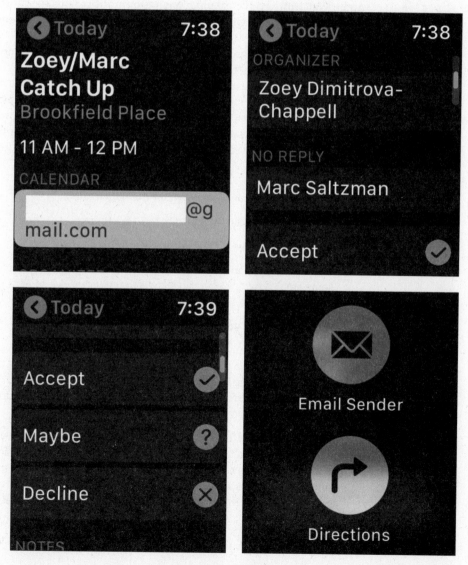

FIGURE 6-18:
Responding to a calendar request is as easy as selecting it, accepting it, and sending the sender a confirmation email.

Don't forget, Apple Watch pulls calendar events from your iPhone. And you don't always have to manually check your calendar for upcoming appointments because you should receive a notification about it (and feel a slight pulse). Some users like to be reminded an hour before an event, for example, whereas others might only want a five-minute reminder. This is all handled in your iPhone's Calendar app.

Accepting a calendar request through notifications

Similar to the step-by-step instructions in the previous section, you can set your Apple Watch to receive calendar invitation notifications — via the Apple Watch app on the iPhone — which you can accept or decline immediately.

The following steps show you how to accept or decline a calendar invitation or to reply to the organizer from your Apple Watch.

As a set up for these steps, assume that someone sent you a calendar invite via email or message. You should receive a notification with the proposed meeting date, time, and information, such as "Natalie's B-Day Party, September 19, 2019, 6 p.m." This is where you have a chance to act on the invite without reaching for your iPhone.

1. **Use your finger to swipe down on the notification.**

   Alternatively, you can twist the Digital Crown button.

2. **Tap Accept, Maybe, or Decline.**

   You can't suggest an alternative date or time or anything, like you can do with some email programs, but this lets the organizer receive some sort of response.

3. **Add the event to your calendar.**

   You don't need your iPhone to receive calendar alerts on your Apple Watch. Therefore, if you go on a run or accidentally leave your iPhone at the office, you can still see existing calendar entries, but you can't add one. Why? Because Apple Watch doesn't have a keyboard, you have to use Siri to dictate new appointments, which isn't possible without the iPhone.

   Alternatively, don't forget that you can raise your wrist or say "Hey Siri" — or press and hold the Digital Crown button — and then say something like, "Add calendar entry, dentist appointment, for 9 a.m. tomorrow." This spoken text is added to your calendar and synced with your iPhone too. See Chapter 7 for more ways to use Siri to help you complete tasks with your Apple Watch.

Setting Reminders on Apple Watch

Introduced in watchOS 6 — the operating system update that debuted in the fall of 2019 — Apple Watch now has an easy way to set reminders and notify you when it's the right time (or place). And, for those who use the Reminders app on iPhone and iPad (running iOS 13 and later) and/or the app on macOS Catalina (or later), the Reminders app on Apple Watch will have a similar look and feel.

To see/set a Reminder on Apple Watch:

1. **To see your Reminders, tap the white Reminders app on the Apple Watch Home screen.**

   Or raise your wrist to activate Siri and say "Open Reminders" or "Do I have any upcoming Reminders?"

2. **Tap one of the three sections to review your Reminders: Today, All, or Scheduled (upcoming Reminders you've set), as shown in Figure 6-19.**

   Of course, you'll be notified of a Reminder at the time you've selected to be reminded, either by a slight vibration on your wrist, alarm sound, or both (you can select in Settings).

REMEMBER

   All your Reminders are conveniently synchronized between your Apple devices — unless you don't want to (which you can tweak in Settings).

FIGURE 6-19: New to Apple Watch is a Reminders app, to help you stay on your game wherever life takes you! Launch the app, review your Reminders, or create a new one.

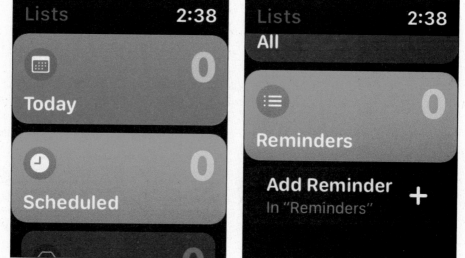

3. **To set a reminder inside the Reminders app, scroll down using your finger on the Apple Watch screen or by turning the Digital Crown and select Add Reminder.**

    You can now create a Reminder by using speech (voice to text) or the Scribble feature (handwriting to text), as shown in Figure 6-20.

If you see the Reminder notification when it arrives, you can swipe the screen to dismiss it or turn the Digital Crown to scroll to the bottom of the reminder, and then tap Snooze, Completed, or Dismiss.

If you discover the notification later: Tap it in your list of notifications, and then scroll and respond.

You can also create Reminders in these alternate ways:

» When in the Reminders app, press firmly on the screen, which immediately launches the Add Reminder option.

» Raise your wrist and ask Siri for a reminder, such as "Tomorrow at 9am, remind me to email my editor" or "When I get to the office, remind me to create new passwords" (the latter example first requires you to add addresses, such as home and office, to your Contacts profile).

Accessing Apple Watch's Integrated Calculator

From Calendar to Calcuator, Apple Watch now includes a handy calculator on your wrist, with the launch of the watchOS 6 operating system in the fall of 2019. While not super high-tech or anything — heck, I had a Casio calculator wristwatch in the '80s! — it sure is convenient to have an integrated calculator app on Apple Watch.

To use the Calendar app, simply tap the Calculator app icon on your Apple Watch Home screen. You can then perform a math equation, including addition, subtraction, division, and multiplication. (See the left side of Figure 6-20.) Press C to Clear the number.

Alternatively, you can ask Siri to open the Calculator app (or just ask Siri your math question if you want to verbalize it).

If you want your Apple Watch to quickly generate a tip if, say, you're at a restaurant, type the amount of the bill (like $120, as shown in Figure 6-20), and it generates the tip for you (default is 15%, but you can turn the Digital Crown for a higher or lower number). You can also select how many people are splitting the bill, so you know how much each person owes!

FIGURE 6-20: The Calculator app for Apple Watch.

Creating and Listening to Voice Memos

Sometimes it's more convenient to simply use your voice than type information.

There is a dedicated Vocie Memos app on the Apple Watch Home screen or you can add a Voice Memo complication (Chapter 4) to your favorite watch face so you can quickly tap and leave yourself a voice recording via the Apple Watch microphone. If wireless earbuds or headphones are connected to Apple Watch, you hear the voice recordings there and not through Apple Watch itself.

Tap the Voice Memos app to open it. Or raise your wrist and instruct Siri to open the app ("Open Voice Memos").You can then perform the following tasks:

>> **Record a voice memo:** Tap the large red Record button, as shown in Figure 6-21, and begin talking into your wrist.

>> **Stop recording a voice memo:** To stop the recording, tap the red Stop button to end.

>> **Play back a previously recorded voice memo:** Tap the name of it below the Red record button. If you didn't rename it, it's listed as Recording 1, Recording 2, and so on. Press Pause to pause the playback, or skip forward or back 15 seconds, as shown in Figure 6-21.

>> **Edit the name of the recording:** Tap it (such as Recording 3) and you have the option to use your voice or Scribble to create a custom name.

» **Erase the voice memo:** Tap the three little dots in the bottom right of the screen to erase the voice recording. You're asked to confirm this is what you want to do.

FIGURE 6-21:
You can now leave yourself handy voice notes through your Apple Watch, and play them back whenever you like.

Recordings are also in Voice Memos on your iPhone, iPad, and Mac.

In fact, if you press and hold on the screen inside the Voice Memos app on Apple Watch, you'll have a choice to play back recordings stored on the Apple Watch itself (section is called Watch Recordings), or tap All Recordings (to hear recordings from other Apple devices, too).

Navigating the Maps App

Apple Maps — or simply Maps — is a standard Apple Watch app that allows you to get directions for the best route from your current location to a destination of your choosing. Because all Apple Watch models have integrated GPS (except for the first model from 2015), the app talks with satellites to help determine your exact location. Think about that for a moment: Your wristwatch is talking to satellites in space. Crazy! Pairing GPS with mapping software means you can easily find your destination or share your location. What's more, Apple Watch Series 5 has a Compass feature, which can also help you navigate accurately. More on this shortly.

When you're en route somewhere, you should see — and feel — the turn-by-turn navigation instructions to guide you along the way, and you can always search for nearby businesses, such as a restaurant or a gas station, simply by asking Siri for it.

To use the Maps app on your Apple Watch, follow these steps:

1. **Press the Digital Crown button to go to the Home screen.**

2. **Tap Maps.**

   This launches the Maps app. Depending on where you left the app last, you see one of two things on your Apple Watch screen:

   - *A menu:* This asks you to search for a location, identify your location, or view your most recent searches (see Figure 6-22). Simply tap the option you want.

   - *An overhead map of your current location:* Here you simply swipe in a given direction to move the map around, or you can twist the Digital Crown button if you want to see nearby streets or businesses.

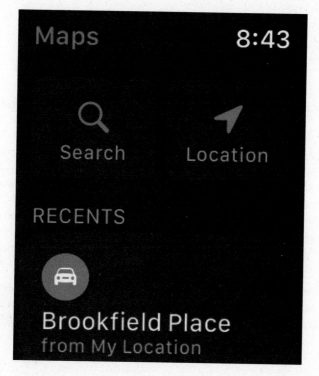

FIGURE 6-22: You may see these options when launching the Maps app on Apple Watch. Choose wisely, Jedi.

3. **In Maps view, tap the blue icon in the lower left of the screen — as shown in Figure 6-23 — to return to your current location.**

   This recenters the map to your specific location. You can also zoom in and out by twisting the Digital Crown button.

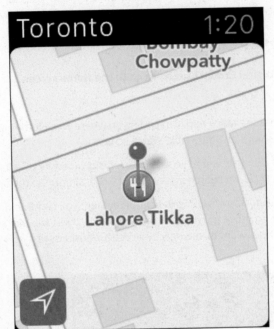

FIGURE 6-23:
The Maps app offers a number of features to help you with directions or find local businesses.

4. **Find a location in Maps view by pressing and holding the screen.**

   You're presented with two options:

   - *Search Here:* Select this to search (Figure 6-24) for something nearby (like a coffee shop or gas station). You're prompted with options. Search by Dictation (use your voice to say what you're looking for and want directions to), use Scribble (write a word using your fingertip), or select Contacts to find someone in your address book. You also have the option of pulling up nearby information divided into the categories: Food, Shopping, Fun, and Travel. Whatever you select, the app pulls up relevant information around your location.

   - *Transit Map:* Pressing and holding this option allows you to view a new map overlaid with bus, train, and/or subway information. You can tap the colored line to pull up additional information, as shown in Figure 6-25.

FIGURE 6-24:
The Map
screen's Search
Here option
(left) allows
you to search
with Dictation,
Scribble, or
Contacts
(middle) when
you're lost,
or to pull
up nearby
information
(right).

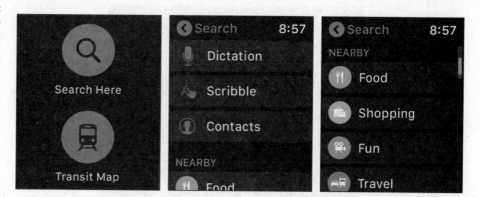

FIGURE 6-25:
Apple Maps
has transit
maps for
bus, train,
and subway
information
(left). Tapping
a transit option
pulls up addi-
tional details
and directions.

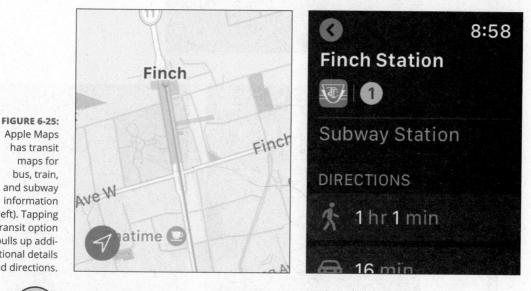

TIP

The next time you search in Maps, you should also see your last searched addresses, as mentioned in Step 2 (see Figure 6-26).

5. **Firmly press on the screen to use Apple Watch's Digital Touch feature.**

If the location is a business, you can tap it on the map to bring up information, such as the address and phone number (which you can call), hours of operation, and its star rating (average user rating out of five stars — via Yelp). You should also see an estimate on how long it might take to get there by foot or by car. Figure 6-27 shows information about a restaurant and how long it might take to travel there.

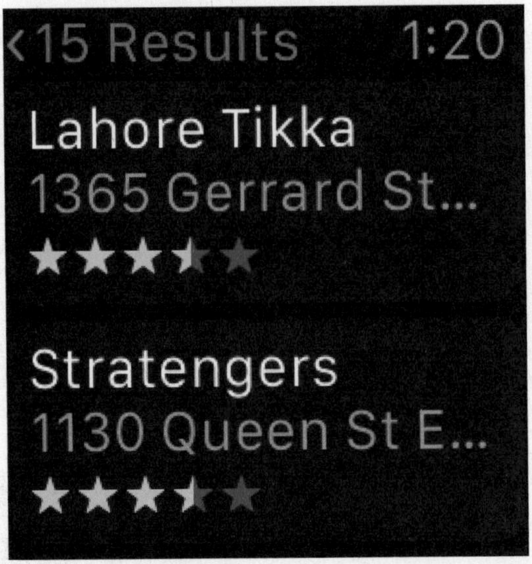

FIGURE 6-26: Ask and ye shall receive. You need to use your voice to get directions or find a local business.

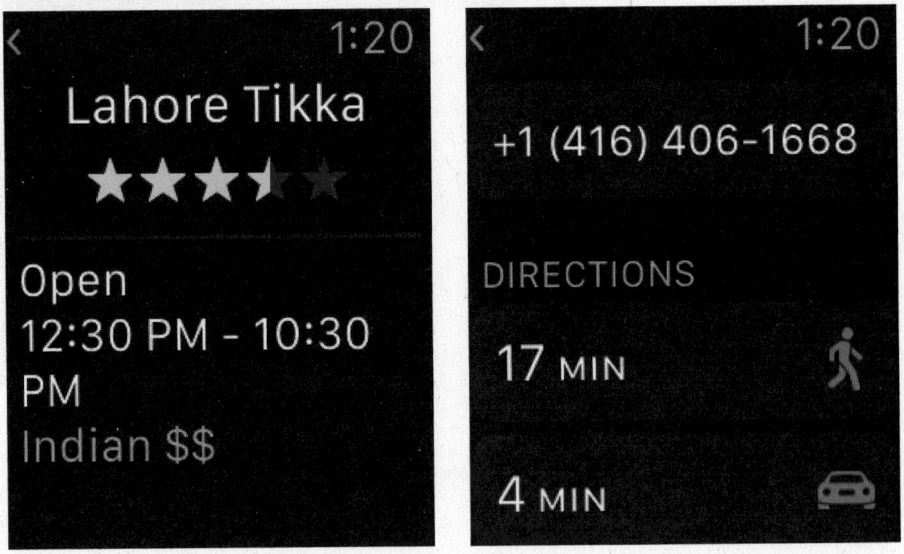

FIGURE 6-27: Get information on a local business — provided by Yelp — simply by tapping its name.

6. **Tap which mode of transportation you're using.**

   You should also see the destination as a pushpin on the map.

   And if you own an Apple Watch Series 5 or later, don't forget to look at the little compass dial in the lower left corner of the Maps screen to see where north is.

TIP

Addresses in Messages, Email, Calendar, and other apps are highlighted in blue and underlined, which means they're tappable. Tapping the address, such as 123 Yonge St., launches that address in the Maps app for you. Neat, huh?

7. **Tap Start to map your route.**

And you're off! Now follow the instructions as you make your way to your destination. If you need to turn right, you feel a steady series of a dozen taps on your wrist at the intersection you're approaching. To turn left, you feel three pairs of two taps. Of course, you can also look down at your screen for visual cues (while on foot and not while driving of course!).

As you can with all other built-in apps, use Siri to access Maps information whenever you like. Either raise your wrist and say "Hey, Siri" followed by something like, "Show me 5 Main Street in Beverly, Kansas" to see it on a map, or "Hey, Siri, give me directions to the Golden Gate Bridge." You can also press the Digital Crown button to activate Siri. See Chapter 7 for more on using Siri to help you perform tasks with your Apple Watch.

Using the Compass

If you own Apple Watch Series 5 or later, there's a built-in compass that points north inside the Maps app to help you get to where you're going.

This is all thanks to an integrated sensor called a magnetometer that detects levels of magnetism.

As shown in Figure 6-28, the little blue compass is in the lower left corner of the screen.

Figure 6-29 shows how you can add a Compass complication to most watch faces on Apple Watch in case you like seeing this information all the time.

You can go into the standalone Compass app for a deeper dive of magnetometer info, displaying your heading along with elevation, latitude, longitude, and incline; see Figure 6-30.

FIGURE 6-28:
The updated
Maps app
now shows
the direction
you're facing.

FIGURE 6-29:
Apple has
added a
Compass
complication
(lower
middle) to
add to several
watch faces,
which shows
directional
information at
a glance.

FIGURE 6-30:
Lost in the woods? You might be just fine thanks to the Compass app on an Apple Watch Series 5.

3

It's All in the Wrist

Gain a helping hand from Siri, which can assist you in completing tasks with your Apple Watch, including tips and tricks to speed up your requests.

Get physical with help from the Activity and Workout apps on your Apple Watch, including setting and modifying goals and earning rewards for your achievements. A new Trends feature shows whether any metric is headed up or down over time, so you can keep it going or turn it around.

Learn all about the Breathe app, how to access the ECG, fall detection, and SOS features, and use the Cycle Tracking app.

Minimize damage to your ears with the integrated Noise app, which monitors the noise level around you.

Access music, podcasts, audiobooks, and radio plays on your Apple Watch and take your favorite media with you anywhere.

Turn your Apple Watch into a virtual wallet by setting up Apple Pay. Then, explore the Wallet app, where you can store movie and concert tickets, boarding passes, coupons, loyalty cards, and much more.

» **Reviewing how Siri works on Apple Watch**

» **Using examples to showcase Siri on Apple Watch**

» **Mastering Siri, including many tips and tricks**

» **Trying out a few extra practical ways to chat with Siri**

» **Having fun with Siri's humor**

Chapter **7**

Siri Supersized: Gaining the Most from Your Personal Assistant

Although many Apple Watch wearers will interact using their fingers on their wrist-mounted gadget — tapping, pressing, or swiping the screen or accessing the two buttons along the side — you can get more done in less time if you simply talk to your watch.

Already an iconic feature on other Apple products (iPhone, iPad, Mac, HomePod, and Apple CarPlay-enabled vehicles) Siri (pronounced "sear-ree") is your own voice-activated personal assistant. Using your words rather than your fingers to ask for information or give a command is a very natural, fast, and simple way to get answers to questions, open up apps, control your smart home, and much more.

After all, talking to your tech gadget is more intuitive than typing or tapping — and getting a humanlike response is more meaningful too — so Apple Watch wearers will no doubt benefit from the fact that the watch has a built-in microphone and speaker.

An Internet connection is required to send your words to Apple's servers for processing. So as long as your watch is on the Internet via Wi-Fi or wirelessly tethered by Bluetooth to your iPhone, or if you own a cellular-supported model, Siri might just be the best feature of your Apple Watch. Of course, you might not always be in a place where you can talk openly (such as in a quiet boardroom meeting) or you might not have Internet access at that moment (such as on an airplane without Wi-Fi or cellular support), but most of the time, you can use Siri to give you what you want — and quickly.

But you don't know where to start, you say? No problem. I cover Siri extensively in this chapter.

Shameless plug alert: As the author of the book *Siri For Dummies,* I show you in this chapter all the different ways you can use Siri to get information on your Apple Watch.

Setting Up Siri on Your Apple Watch

You have four different ways to call up your personal assistant on Apple Watch:

>> Press the Digital Crown.

>> Say "Hey Siri."

>> Raise your hand towards your mouth.

>> Select the Siri watch face (see Chapter 4 for info), add the Siri complication, then tap it to talk to Siri.

Cool, huh?

First, the good news: You probably already set up Siri on your iPhone when you first turned on your device. This gives you access to your assistant on Apple Watch too. As you may or may not recall from previous chapters, your iPhone asked you if you want Siri (and, yes, you can also enable or disable it in the Settings ⇨ Siri & Search area of your iPhone). See Figure 7-1 for a look at the Siri options on iPhone.

Selecting a language

When you set up your iPhone, it also asked for your language preference. You have nearly 40 options, including English, French, Spanish, Italian, German, Chinese, Japanese, Korean, and Russian — to name a few. Choosing a language and dialect isn't just so Siri can speak in a language you understand; it's also to let your new personal assistant better understand you. For example, someone from the United States or Canada will say "Call Mom" differently than an English-speaking person from the United Kingdom or Australia. One might sound more like "Coll mum" or "Cull mam" and so on.

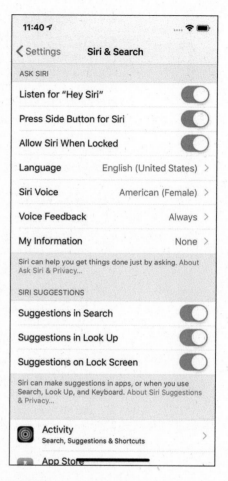

FIGURE 7-1:
Select your Siri options on your iPhone, such as voice gender and language.

In fact, you can choose from nine different kinds of English for Siri: Australia, Canada, India, Ireland, New Zealand, Singapore, South Africa, United Kingdom, and United States (notice how Scotland isn't in there? Siri isn't THAT good!). Obviously, Americans have various accents too — differences definitely exist between speakers from Long Island, Boston, Dallas, and Minneapolis, for example — but American English can be vastly different from the English spoken in London or Sydney. Thus, be sure to choose the correct language from the list or you may have some difficulties understanding Siri — and vice versa.

Choosing a Siri voice and setting up search options

It's also important to note Siri has a female voice in the United States by default, but you can change it to a male voice if you like. For this reason, I usually refer to Siri as "it" to keep language universal.

To make sure Siri works flawlessly on Apple Watch, you need to first go to Settings ⇨ Siri & Search on your iPhone and flick the toggle to green ("On") for the following three features:

>> **Listen for "Hey Siri":** You can decide whether you simply want to say "Hey Siri" instead of pressing the Home button on an iPhone or pressing the Digital Crown button on Apple Watch. If you enable this feature in your iPhone's settings, you can simply raise your wrist to your mouth and say "Hey Siri," followed by your question or command.

>> **Press side button for Siri (iPhone X and newer):** Make sure this is also turned on (flicked to green). This allows you to press the Digital Crown on your Apple Watch to initiate your personal assistant. (On iPhone 8 and older, you need to press and hold the Home button to activate Siri.)

>> **Enable "Raise to Speak":** If you like the idea of simply raising your wrist to wake up your personal assistant, on your Apple Watch, tap Settings ⇨ General ⇨ Siri. Choose whether you want to turn Raise to Speak on or off. You can also enable to disable the "Hey Siri" verbal option here, too, instead of on your iPhone.

Other options you have inside the Siri & Search settings on your iPhone:

>> **Voice Feedback:** By default you will hear Siri talk through the watch, but you can make a few tweaks.

   ● *Control with Ring Switch:* This silences all audio feedback, but you'll continue to hear Siri beep and provide voice feedback when you say "Hey Siri."

   ● *Hands-Free Only:* Siri only beeps and provides voice feedback when you are using "Hey Siri" or when you connect to a Bluetooth device, headphones, or CarPlay.

>> **My Information:** Selecting this tab opens your Contacts app. Why? Your iPhone wants you to point Siri to your own name to learn where you live and who is close to you. If you don't have an entry in your Contacts with your own information, you need to create one. Once you fill out all the fields, you can

tell Siri "Take me home." Of course, Siri won't know where that is unless your address is listed on your Contacts page. Similarly, you can tell Siri to call your wife or husband or email your dad or text your mom — all of which you can identify in your iPhone's Contacts page (yes, fields exist there for people close to you).

You can activate Siri on your Apple Watch or iPhone and tell it to call you something else. Instead of Robert, you can say "Siri, call me Bob" or "Siri, call me Junior" for a nickname. Going forward, Siri addresses you by the name you prefer. You can always change it if you like.

Unlike other speech-to-text technology, including those offered by other smartphones, Siri works on the *operating system* level and knows which app to open based on your request. Most other smartphones require you to first open an app *before* you tap the microphone to speak. You don't need to do this for Siri.

Ready to rock? Connecting and talking to Siri

Before you get started, you need to ensure you have a good Internet connection. As I cover in this chapter, you need to be online for Siri to work. You can see the signal strength of Wi-Fi and cellular on your Apple Watch; without a good connection, you might find Siri inaccessible.

Okay, to connect to Siri:

1. **Make sure your Apple Watch is connected to the Internet by cellular, Wi-Fi, or Bluetooth (and/or connected to your nearby iPhone).**

2. **On your paired iPhone, tap Settings ⇨ Siri & Search.**

   Make sure that the Listen for "Hey Siri" option is on. On iPhone X or later, make sure that the "Press Side Button for Siri" option is on. On iPhone 8 or earlier, make sure that the "Press Home for Siri" option is on.

3. **On your Apple Watch, tap Settings ⇨ General ⇨ Siri.**

   Choose whether you want to turn Hey Siri and Raise to Speak on or off. (See the previous section for more information on these options.)

4. **On Apple Watch Series 3 or later, scroll down to Voice Feedback, then choose when Siri speaks.**

   Options include: Always On, Control with Silent Mode, or Headphones Only.

Talking to Siri on Apple Watch

Before you start talking, remember you can maximize Siri's performance on Apple Watch by following these tips:

>> **Speak clearly:** I know this can be difficult to be conscious of, but the less you mumble and the more you articulate your words, the better Siri works. Don't worry: Siri is remarkably keen on picking up what you say (and even what you mean), so you don't need to speak like a robot. Just be aware that you'll get better results with clearer speech.

>> **Find a quiet place:** A lot of background noise isn't great for Siri because it might not be able to pick up what you're saying very well. The quieter the environment, the better Siri can understand your instructions. This might be tough if you're in a crowded restaurant, driving with the window open, or walking down a busy street, of course, so you might need to speak a little louder and closer to the Apple Watch microphone.

Now, when you ask Siri a question — such as "What's the weather like in Seattle tomorrow?" — you should see colors dance around the bottom of the watch screen to confirm it's listening to you. Stop talking after you're done and you should hear a beep to confirm Siri is now processing your request.

If you make a mistake while asking Siri a question (maybe you accidentally said the wrong person's name to text) or perhaps Siri didn't hear you clearly, you can tap the screen to nullify the request and then ask again. You should hear the familiar *ping* tone to confirm Siri is listening for your new request.

The final thing you should see is when Siri performs your desired action. Siri might open up a map, an email message, a calendar entry, a restaurant listing, or show you such information as the score of your favorite team's last game (without even opening an app). Depending on what you ask Siri, you may see — rather than hear — the information. For example, if it's a dictionary definition or a numerical equation you're after, you might hear something like "Here you go" or "This might answer your question" and then Siri shows you the information on the screen. Other times it'll tell and show you the answer.

TECHNICAL STUFF

Because all requests to Siri are uploaded to a server, it's not unheard of for the server to be temporarily inaccessible — but it doesn't happen very often. Siri will apologize to you and ask that you please try again later. A problem with Siri *isn't* an indication of a problem with your Apple Watch or iPhone, so don't fret. The outage is usually only a couple minutes (if that), but it's something you should be aware of.

What Are Siri Shortcuts?

With iOS 12 or later, Siri Shortcuts let you quickly perform everyday tasks with the apps you use the most by simply asking for Siri or tapping your Apple Watch or iPhone. You can choose if this works on your iPhone's Lock screen too.

Basically, Siri learns your routines across your favorite apps and then your assistant suggests an easy way to perform common tasks. For example, if you typically ask for weather at the same time of day, Siri may prompt you with the info from your favorite weather app. If you like to order a coffee to pick up every morning with the same app, Siri might suggest that beverage you tend to pick.

Now, that's smart!

To use a Siri Suggestion, just tap it on the Lock screen, or swipe down from the center of your screen to show Search, then tap the Siri Suggestion.

You can also add shortcuts to Siri. Look for the Add to Siri button in your favorite apps — with hundreds already supported — and then tap to add with your own personal phrase. Or go to Settings to find all shortcuts available on your device.

Using the Siri watch face and Siri Shortcuts

As I cover in Chapter 4, you can choose from many different watch faces for your Apple Watch. The aptly named Siri watch face is tied to your personal assistant, updates throughout the day, and shows relevant content and information you might need based on your location, time of day, and routines. For example, you might see calendar events, boarding passes, or your favorites from the Home app (including controlling and monitoring your smart home devices). And the Siri watch face supports Siri Shortcuts. See Figure 7-2 for a look at Siri in action.

To use the Siri watch face:

1. **Press and hold the watch face on your Apple Watch.**

2. **Swipe left or right until you land on Siri.**

3. **Let go of the screen and the Siri watch face will be set.**

FIGURE 7-2:
The Siri watch face can show you relevant information tied to when it is and where you are, plus it supports Siri Shortcuts to initiate requests you typically call for.

Using Siri effectively on Apple Watch

Everything you can do with your fingers, you can do with Siri — if not more — and in less time. A good way to demonstrate its versatility is to look at a number of built-in Apple Watch apps and some examples of how you can use Siri to get what you want. See Figure 7-3.

Siri works in the main Home screen mode, while accessing the clock, or in any app you find yourself in, such as asking Siri for map directions even though you're using the Music app at the time.

TIP

If you're unsure about all the things Siri is capable of, say "Siri, what can you do?" on your iPhone or Apple Watch and you should see a huge list of things!

Clock/World Clock apps

Some examples of using Siri for time-related tasks include:

>> "What time is it?"

>> "What time is it in Dubai?" Figure 7-4 shows what you might see after asking this.

FIGURE 7-3:
When setting up Apple Watch via the iPhone app, enable Siri. Then, on Apple Watch in the Settings, allow for "Hey Siri" to activate your hands-free personal assistant.

>> "What time will the sun rise in Brisbane?"

>> "How many days until Christmas?"

>> "What day of the week will it be on June 10, 2020?"

> **Dubai, the Unit-ed Arab Emirates**
> Current Time
>
> **9:09**PM
>
> +8HRS

Messages app

Some examples of using Siri for sending and receiving messages include:

>> "Read me my messages."

>> "Do I have any messages from ___ (name)?"

>> "Text my wife 'Hey, hon, how's your day going?'"

>> "Text 212-555-1212. 'I'm looking forward to our after-work drink tonight.'"

>> "Text Julie and Frank 'Where are you guys?'"

>> "Reply 'That's awesome news!'"

TIP

Something fun to try with Siri on Apple Watch — which may amuse the kids — is to ask Siri to remind you about something really far in the future. For example, I asked Siri to remind me to kiss my wife in 10,000 years, and I was asked if it should be placed in my calendar then. (I hope someone finds a cure for mortality soon!)

Phone/Contacts apps

Some examples of using Siri for making calls or looking up Contacts information include:

>> "Call Mom."

>> "Dial 212-555-1212."

>> "What's Michael Smith's address?"

>> "What's my sister's work address?"

>> "Learn how to pronounce my name."

>> "Show Jennifer's location."

Mail app

Some examples of using Siri for looking for email include:

>> "Show me my email" or "Check email."

>> "Do I have any email from David Smith?" Figure 7-5 illustrates what happens when you ask for email from a specific person or company. Swipe down to see the correspondence between you and this person.

>> "Show the email from Natasha yesterday."

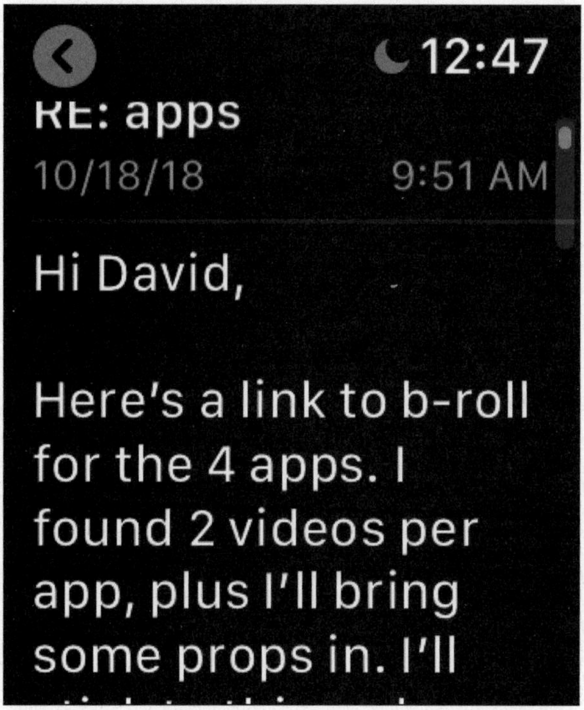

FIGURE 7-5:
Ask for mail and Siri shows it to you — whether you want to view it by time or person/company.

REMEMBER

You can't compose a brand-new email on your Apple Watch, but you can tap a message and reply to one. Also be aware, when you ask Siri to show you email on Apple Watch, it may suggest you look at your phone if it doesn't support the email client you use.

Calendar app

Some examples of using Siri for accessing calendar information include:

>> "Show me what appointments I have on Monday."

>> "When is my next meeting?"

>> "When's my next appointment?"

>> "Move my 12 p.m. meeting to 1 p.m."

>> "Cancel the meeting at 4 p.m."

Activity/Workout apps

Some examples of using Siri for fitness-related tasks include:

>> "Open the Activity app."

>> "Open the Workout app."

>> "See Move information in Activity."

>> "See Stand information in Activity."

>> "See Exercise information in Activity."

>> "Open Indoor Walk in Workout app."

>> "What's my heart rate?"

Maps app

Some examples of using Siri for looking for directions or for a local business include:

>> "Show my location on a map."

>> "Where is my closest coffee shop?"

» "Take me home." Figure 7-6 shows a sample result. Siri might ask you where you live the first time you say this (or it pulls the information from your Contacts).

» "Take me to Grand Central Station."

» "What's my next turn?"

» "Give me directions to Mom's office."

» "Find a gas station."

» "Find the best sushi restaurant in Miami."

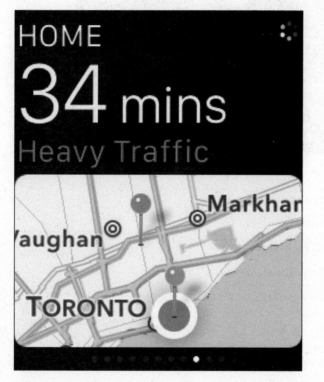

FIGURE 7-6:
Use your voice in the Maps app — perhaps to take you home.

Music app

Some examples of using Siri for playing music include:

» "Play Madonna." (See Figure 7-7.)

» "Play Workout playlist."

FIGURE 7-7:
Ask Siri to play music — whether it's an artist, song, album, genre, playlist, and so on.

>> "What song is this?"

>> "Shuffle my music."

>> "Play rock music."

>> "Play 'Sugar' by Maroon 5."

>> "Skip this track."

Speaking of music, introduced to Apple Watch in late 2019 is the ability to "Shazam" a song — that is, finding out the name of the track and who sings it — even over cellular connectivity if you don't have your iPhone nearby.

As shown in Figure 7-8, when you hear a song that catches your ear you can raise your wrist and ask what it is.

What song is this

Listening...

Sounds like you're listening to 'Talk' by Khalid.

FIGURE 7-8: You can now ask Siri to identify the song for you, with the help of Shazam.

Web searches

It's easier than ever to use Siri for a web search on your Apple Watch. The results appear on screen for you to read and scroll through, and you can tap a URL to take you to the appropriate page. As shown in Figure 7-9, Siri is more helpful than ever before — including the scouring of web pages for you.

Miscellaneous

Some other examples of a few other random — but fun — things you can do with Siri on Apple Watch include:

>> "What's the weather outside?"

>> "What's it going to be like this week?"

>> "Do I need a coat?"

>> "How are my stocks doing?"

>> "What's the Apple stock at?"

>> "How's the Dow Jones doing?"

What is a good marathon training plan for beginners
Here's what I found:

How to run a marathon — free marathon training
www.runnersworld.-com

Jan 1, 2019... Training for a marathon?

FIGURE 7-9:
Ask and ye shall receive! Apple Watch now displays web results tied to Siri queries.

>> "Set an alarm for 7 a.m."

>> "Set an alarm for one hour from now."

>> "Wake me up in 30 minutes."

>> "Open Passbook."

>> "Open Stopwatch."

>> "Open Timer."

>> "Open Settings."

Because it takes only a quick Siri request to set up a reminder, you might be tempted to do this while driving. But even a minor distraction could cause an accident, so resist using Apple Watch and Siri until you've parked the car.

Trying Other Tasks with Siri

Siri is one of the fastest, easiest, and most accurate ways to interact with content on your Apple Watch. But Siri can perform lesser-known yet impressive feats too, and the following sections discuss some of my favorites. Just start your request with "Hey Siri" and Siri will ask how it can assist you, as shown in Figure 7-10.

Setting reminders by location

It's a breeze to ask Siri to remind you of something by time — such as "Tomorrow at 10 a.m., remind me to call the dentist to book an appointment" — but did you know you can set reminders by location too?

For example, raise your wrist and say "Hey Siri, remind me to call Mom when I leave here." Whenever you leave wherever you are — such as your office,

FIGURE 7-10:
Siri can assist you with some lesser-known — but impressive — feats.

a coffee shop, or a shopping mall — Siri reminds you to call your mom. Your nearby iPhone's integrated GPS means it's location-aware.

Another example: Say "Remind me to take out the trash when I get home." Because Siri accesses your home address from your iPhone's Contacts app, you won't be reminded of the chore until you pull into the driveway.

REMEMBER

If you haven't added your home address information yet, see the "Setting Up Siri on Your Apple Watch" section for how to do this.

Reading your texts

Many Siri users are aware you can dictate your text messages; simply say something like "Hey Siri, text Mary Smith 'Please don't forget to call the florist for tomorrow's event.'"

But did you know you can have your text messages read to you? Raise your wrist or press the Digital Crown button and then say something like "Read my texts." Once Siri reads a message to you, you can say something like "Reply saying 'That's an excellent idea — thanks'" or "Tell her I'll be there in 20 minutes."

You can also ask Siri something like "Do I have any texts from Mary?"

Calculating numbers

Siri includes support from Wolfram Alpha's vast database of facts, definitions, and even pop culture information. (For example, ask Siri who shot J.R. or Mr. Burns!) But you can also ask Siri to perform math problems for you. If you're adding up checks to deposit at the bank, for example, ask Siri something like "What's $140.40 plus $245.12 plus $742.30 plus $472.90?" and within a moment, you should hear the correct answer (which is $1,600.72).

If you're out with friends at a restaurant and the bill comes to, say, $200, you can also ask Siri something like "What's an 18 percent tip on $200?" and Siri tells you how much that is ($36).

Of course, Siri can also handle multiplication, subtraction, equations, fractions, and more. All you have to do is ask!

Finding your friends

If you're not familiar with the free Find My Friends app on Apple Watch or iPhone, it uses GPS to provide your geographical location to people you choose to share this information with, such as a spouse, kids, grandkids, friends, or coworkers. Once you add consensual people, you can also see their whereabouts on a map — represented by colored orbs — and get the address they're at if desired.

You probably saw this one coming: You can use Siri to get the most from the Find My Friends app. Raise your wrist and say "Hey Siri, where are my friends?" This opens the Find My Friends app, and you should see who's around and how far they are from you. Now you can send someone a message, such as "Let's grab a coffee" on your phone. You can also ask Siri something like "Is my husband at home?" or "Where's John Smith?" or "Find my sister."

To get started, go to the Find My Friends app on your iPhone to set things up on your Apple Watch and to add friends; you can't do this from Apple Watch. You can then enjoy this feature because all models have integrated GPS (except the first Apple Watch in 2015).

Extending the Fun (and Silly) Ways to Interact with Siri

Siri is pretty funny — if you haven't yet figured this out from talking to it on your iPhone. In case you haven't, the following are some fun and cheeky things you can ask Siri for on your Apple Watch — and the kinds of responses you can expect.

Spoiler alert: Only read the **bold** questions and not the answers if you want to see what Siri replies with on your own!

Say: "What's the best smartwatch?"

You don't expect Siri to recommend a rival Android-powered watch, do you? Instead, it answers this question with "The Apple Watch will show you a really good time" or "I say Apple Watch — hands down" or another answer.

Say: "I love you, Siri."

Deep down, Siri might be flattered, but it suggests otherwise. Siri might write something like "You hardly know me" or "That's nice — can we get back to work now?" or "Impossible!"

Say: "Siri, I'm bored."

If you find yourself bored while wearing your Apple Watch, you can tell Siri how you're feeling and it replies with something like "Not with me, I hope." Or it'll converse with you — be it offering a story, song lyrics, or a poem or engaging in a "knock, knock" exchange if it's in the mood.

Say: "Who's your daddy?"

This one borders on the naughty. While Siri was a little reluctant at first, it knows which side its bread is buttered. You might hear "You are" or perhaps something like "I know this must mean something — everybody keeps asking me this question."

Say: "What's the meaning of life?"

You can ask Siri a profound question, such as "What's the meaning of life?" and although it might give you a literal translation, you might also see a reply with something cheeky, such as "A movie" or "All evidence to date suggests it's choco-late." Or "I don't know, but I think there's an app for that."

Say: "Will you marry me?"

After professing my affection for Siri (it writes "That's sweet," "I sure have received a lot of marriage proposals lately," or "You are the wind beneath my wings"), I went for it and asked for Siri's, uh, hand in marriage. Its reply: "Let's just be friends, okay?"

IN THIS CHAPTER

» **Mastering the Activity and Workout apps**

» **Understanding personalized reminders, feedback, and rewards**

» **Using the Breathe app for moments of relaxation**

» **Managing the Cycle Tracking app to help women with their menstrual cycle**

» **Opening your eyes to the Noise app**

» **Using Apple Watch's heart rate sensor, electrocardiogram (ECG) feature, Fall Detection, Emergency SOS, and Medical ID**

Chapter **8**

Fitness Fun and Happy Health: Apple Watch Is Your Workout Buddy and Digital Doctor

Fitness is one of the smartest applications on your smartwatch. Whether you're trying to monitor your regular daily activity, determined to lose weight, an athlete looking to maximize your training, or simply wanting to manage your workout regimes in an easy way, Apple Watch can handle it all. While you wear one of these high-tech yet water- and sweat-resistant devices on your wrist, you'll receive real-time information, such as total distance traveled, calories burned, and other details. Apple Watch also serves as a digital doctor, of sorts, with a bevy of health–related applications on your wrist. And while Apple Watch can show your activity information on its small screen, you can dive deeper and track more details and historical information via apps on your iPhone because it syncs with the same apps.

And be sure to have the latest operating system installed (watchOS 6, at the time of writing this) to take advantage of the latest health and fitness features.

Tracking Your Fitness with Apple Watch

But how does this work, you ask? Apple Watch has a list of features and some powerful technology behind it all:

>> **Accelerometer:** Like other activity trackers, Apple Watch has a built-in accelerometer to count your number of steps, like an old-fashioned pedometer.

>> **Activity trends:** Apple Watch can now send activity trends data to your iPhone, which tracks and compares your progress over the last 90 days to gauge if your fitness levels are improving.

>> **Altimeter/barometer sensor and GPS:** Its built-in altimeter calculates the number of stairs you climb. Its integrated GPS chip tracks how far you've moved.

>> **Breathe app:** For the mindful, Apple Watch includes an app that makes you take a quick break from your day to focus on your breathing.

TIP

The first-generation Apple Watch from 2015 doesn't support these fitness- and health-related features.

>> **Heart rate sensor:** This integrated feature tracks your workout's intensity and, if used every day, might detect if something is off. It records unusually high or low heart rates, and alerts you about them even when you don't feel symptoms.

>> **Fall detection:** Beginning with Apple Watch Series 4, Apple Watch includes an accelerometer and gyroscope to detect when you've fallen. You can initiate a 9-1-1 call or, if you're unresponsive after 60 seconds, the watch automatically places an emergency call and sends your location to your emergency contacts.

>> **Emergency SOS:** This lets you easily call emergency services. You can also notify your emergency contacts, send your current location, and display your Medical ID badge on the screen for emergency personnel. It now works overseas, too!

>> **Noise app:** Apple Watch can help protect your ears by listening to nearby noise levels and the duration of exposure.

>> **Cycle tracking:** Designed to help women manage their monthly menstrual cycle, this easy to use and discrete app shows relevant cycle information.

This chapter covers all this, and more! But right now, I start off this chapter with how to use Apple Watch for fitness using two of its main apps: Activity and Workout.

Getting Up and Running with the Activity App

As Apple explains on its website, fitness isn't "just about running, biking, or hitting the gym. It's also about being active throughout the day."

 Thus, one of the two main fitness apps on Apple Watch is devoted to your general activity levels during a regular day. This includes such things as walking the dog, chasing after your kids or grandkids, and taking the stairs rather than an escalator or elevator. The aptly named Activity app's main screen (shown in Figure 8-1) keeps track of everything physical you do — and encourages you to keep moving.

FIGURE 8-1:
The Activity app shows you multicolored rings based on your movement, exercise, and more.

Quite simply, the Activity app gives you a visual snapshot of your daily activity. It's broken down into three colored rings:

>> **Move:** The reddish-pink ring shows how many calories you've burned by moving.

>> **Exercise:** The lime-green ring is for the minutes of brisk activity you've completed that day.

>> **Stand:** The baby-blue ring gives you a visual indication of how often you've stood up after sitting or reclining.

Your goal is to complete each ring each day by reaching the suggested amount of exercise, as outlined in this chapter. The more solid each ring is, the better you're doing.

The first (main) screen of the Activity app gives you a Move, Exercise, and Stand summary, but if you swipe to the left, you can access a dedicated screen for each of the Activity meters.

As shown in Figure 8-2, you should see a summary of each Activity section, which explains what you're seeing in this app.

Before you begin any Activity, however, Apple Watch wants to learn a little about you first — namely, your gender, age, height, and weight — in order for the numbers to be accurate, such as estimating your calories burned. For example, a 25-year-old female burns calories at a different rate than a 65-year-old male.

You only have to do this once, and you can answer the required questions with your fingertip (see Figure 8-3).

If you live in the United States, you should see customary unit measurements, such as pounds, but those who live in Canada or the United Kingdom fill out information using the metric system. (Ditto for setting distance, be it in miles or in kilometers.)

Move

Moving is good — even if it's not that fast. Motion helps you burn calories and gets your heart pumping and your blood flowing. The Activity app's Move ring tells you how well you're doing based on your personal active calorie-burn goal for the day, as shown in Figure 8-4. In this example, the default goal is 450 calories per day, which is a couple hours of walking around a shopping mall. If that's too easy to reach or, on the flipside, too ambitious, you can easily make necessary adjustments to suit your needs. Just press firmly on the Apple Watch screen (Force Touch) and change the Move goal to something more achievable: Press + or − until you see your desired goal.

FIGURE 8-2: Apple Watch explains how Move, Exercise, and Stand work. You can also share your Activity info, via the iPhone app, which might motivate you to do better!

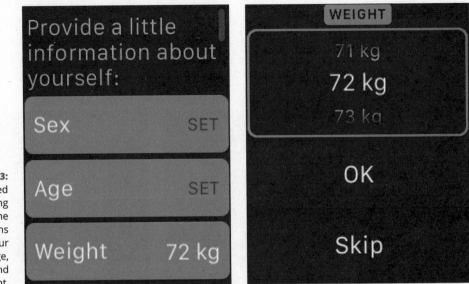

FIGURE 8-3:
Get started by answering some questions about your gender, age, height, and weight.

FIGURE 8-4:
Part of the Activity app, this ring summary screen shows your Move, Exercise, and Stand stats for the day. Just swipe up from the rings screen.

To use the Move tab in the Activity app, follow these steps:

1. **Press the Digital Crown button to go to the Home screen.**

2. **Tap the Activity icon.**

   You can also raise your wrist and say "Hey Siri, Activity." Either action launches the Activity app. You'll see the rings that summarize your fitness goals for the day.

3. **Swipe up on the Activity app's main (summary) screen to see additional info.**

   Move tells you how much you've moved during the day. The red number at the top of the screen is your current estimated calories burned, and how far you are to your daily goal (such as 240/320 CAL). You'll also see it written in a percentage of your daily goal, such as 70%.

   You can change your caloric goal in the Activity app by pressing the watch screen (Digital Touch) and selecting a new goal. Press + or – to set your desired goal. You can also change your Exercise and Stand goals in the same fashion (see Figure 8-5).

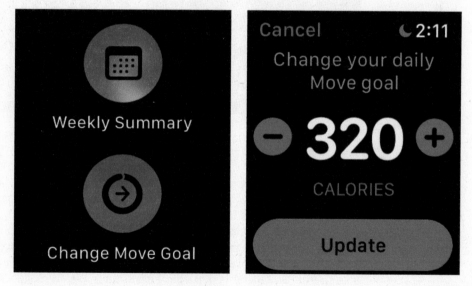

FIGURE 8-5: If you don't like the suggested (default) goals for Move, Exercise, and Stand, simply press and hold the screen in the Activity app (left) to make changes (right).

4. **Swipe up on the screen to see a History graph with each hour of the day presented and how well you've done per hour (highlighted by a vertical line).**

The taller the pinkish bar, the more you moved that hour, as shown on the left side of Figure 8-6.

5. **Swipe up again or twist the Digital Crown button to see even more details, such as total steps, total distance, and flights climbed.**

You can see this on the right side of Figure 8-6.

FIGURE 8-6:
Scroll up within the Activity app to see another view of your day's performance — divided by hour.

You can now go about your business.

REMEMBER

Many different kinds of bands are available for Apple Watch, and active types might prefer the aptly named Sport band made from fluoroelastomer (synthetic rubber), which comes in several different colors. Compared with leather, the link bracelet, and the Milanese loop, the high-performance and smooth Sport band might be most ideal for those who exercise because of its light weight, durability, and resistance to sweat and rain. It also features a pin-and-tuck enclosure for a secure fit. This band is available for all Apple Watch models. See apple.com/applewatch for more on band selection.

Exercise

Whether you want to do something active in one shot — such as jogging on the treadmill after work — or a little bit here and there, it's recommended you do at least 30 minutes of exercise each day. What constitutes "exercise," you ask? How is this different from mere "moving"? Any activity at the level of a brisk walk or above is considered exercise, says Apple. By seeing how much you're exercising — or, rather, not exercising — you might just be motivated to improve your overall health (which has also been proven to be linked to happiness).

To use the Exercise tab in the Activity app, follow these steps:

1. **Press the Digital Crown button to go to the Home screen.**

2. **Tap the Activity app.**

   You can also raise your wrist and say "Hey Siri, Activity." Either action opens the Activity app, taking you to the main Activity screen with a summary of all three rings.

3. **To go to the Exercise area, swipe twice to the left; or, at any time, and regardless of the app you're in, tell Siri "Show me Exercise information" to go right to this screen.**

   You should see a green number in the middle of the screen. This is the total exercise time calculated for the day so far. You will see how close you are to your daily goal, too (such as 25 out of 40 minutes), and listed as a percentage, such as 65%. To change your goals, press firmly on the screen (Force Touch) and tap + or – to set your desired goal.

4. **Swipe up for your History graph, which shows your hourly activity level — measured in minutes — for when you were most active.**

   As you might expect, the higher the line on the graph, the better. Even if you exercise a little here and a little there, every bit helps and goes toward your daily time goal.

5. **Swipe up again or twist the Digital Crown button for additional exercise information on your Apple Watch.**

   You see a numerical summary of your day's achievements. Alternatively, you can grab your iPhone and open the Activity app.

6. **Press the Digital Crown button to return to the Home screen.**

   Don't be discouraged if you're not reaching your exercise goals. Try again tomorrow or reduce the number of your suggested active minutes — from say, 30 minutes to 20 minutes.

Stand

Many of us — including yours truly — have jobs where we sit for a good chunk of the day. This isn't doing much for those love handles we're trying to get rid of. Apple Watch knows when you stand and move around at least for one minute, and this all goes toward your Stand ring. Even if it's just to get up from your computer and go get a glass of water, a small stretch, or a walk down the hall to say hello to a coworker, all this Stand time adds up. You've completed the default Stand ring requirements if you move at least one minute in 12 different hours during the day. The app will remind you to get up if you've been idle too long (about an hour).

(If you want to disable these reminders, open the Apple Watch app on iPhone, go to My Watch, and then Activity. Toggle the "Stand Reminder" setting to the Off position, if desired. Your iPhone syncs with Apple Watch, and stop reminding you to get up.)

To use the Stand tab in the Activity app, follow these steps:

1. **Press the Digital Crown button to go to the Home screen.**

2. **Tap the Activity app.**

   You can also raise your wrist and say "Hey Siri, Activity." Either action opens the Activity app, taking you to the main Activity screen with a summary of all three rings.

3. **The blue ring represents your Stand info, per day.**

   Swipe up to see blue info near the bottom of the screen. This shows how many hours you've stood up for (at least one minute per hour). The second number is the total goal hours (such as 12). You'll also see percentage of your goal for the day.

TIP

Regardless of which Activity screen you're in — the summary page, Move, Exercise, or Stand — you always see a clock in the top-right corner, so you always know the current time without having to leave the Activity app. Smart, no?

4. **Swipe up to see the History graph.**

You should see the day laid out chronologically and a full vertical bar for any hour you stood (for at least a minute per hour).

It doesn't matter if it's consecutive hours or spread out throughout the day. The idea is to get up at least once per hour during the day (unless you sleepwalk, which means you won't have to worry about this overnight!).

5. **Swipe up again or twist the Digital Crown button to obtain more information on your activity.**

For example, you can get a numeric summary of your day's progress, including total steps, total distance, and more. (See the "Understanding the Workout App" section for more.)

Be proud! You're getting your move on.

Understanding the Workout App

You might be wondering how the Workout app differs from the Activity app. Aren't they the same thing?

Not exactly.

Whereas both are fitness related, the Workout app differs from the Activity app in one respect: Instead of showing your progress over the past day, Workout provides real-time information about calories burned, elapsed time, distance, speed, and pace for your walks, jogs, runs, cycling, and indoor equipment, such as an elliptical, a stair stepper, a rower, a treadmill, and more. In other words, rather than generic daily stats, the Workout app shows you cardio information based on what you're doing and while you're doing it.

Although most activity trackers and smartwatches can spit out generic information on your estimated calories burned — by simply moving — Apple's technology is tailored to specific exercise equipment and/or exercises. All you have to do is choose the type of workout you'd like to tackle, and Apple Watch turns on the appropriate sensors, such as accelerometer and gyroscope (for motion), heart rate monitor, and GPS (if exercising outdoors, for example, because GPS requires line-of-sight with satellites above the earth).

You can then receive a detailed summary of your exercise — and, of course, your workout counts toward your Activity ring measurements for the day.

You can also set goals, chart your progress, and earn awards.

REMEMBER

Not including the first-generation Apple Watch, the Workout app can track swimming, too! Don't try this with other fitness trackers or smartwatches, unless you know for certain they're waterproof, too!

To use the Workout app on your Apple Watch, follow these steps:

1. **Press the Digital Crown button to go to the Home screen.**

2. **Tap the Workout app.**

   You can also just raise your wrist and say "Hey Siri, Workout." Either action launches the Workout app, where you see the main Workout screen. You should see options for many different kinds of indoor and outdoor exercises, as shown in Table 8-1.

TABLE 8-1

Different Workout App Exercise Options

Outdoor Walk	High intensity interval training
Outdoor Run	Hiking
Outdoor Cycle	Rower
Indoor Walk	Stair stepper
Indoor Run	Yoga
Indoor Cycle	Swimming
Elliptical	Wheelchair
Other	

Most of the exercise options in Table 8-1 are self-explanatory. The exceptions are the following:

- *High intensity interval training (also known as HIIT):* This is an exercise where you alternate intense movement with rest. So for example you might jump rope for 45 seconds, rest for 30 seconds, and then repeat this with another exercise.

- *Other:* For this option, the app recognizes that you might not find a matching workout type for your activity; you'll still earn the calorie or kilojoule equivalent of a brisk walk anytime sensor readings are unavailable.

3. **Swipe up or down to view a workout, or you can cycle through them by twisting the Digital Crown button forward or backward.**

   What are you in the mood for? Figure 8-7 gives you some of your options.

FIGURE 8-7: Choose a workout and tap it to set a goal. Apple.com gives a quick look — with icons — of many Workout options with some categories having both indoor and outdoor measurements.

4. **When you see a Workout you like, such as Outdoor Cycle, tap the three little dots in the top-right corner to see and set goals.**

   On this screen, you can select a goal:

   - Calories (highlighted in pink)

   - Time (highlighted in yellow)

   - Distance (highlighted in blue)

5. **Use your fingertip to press + or – for the numbers to go up or down, respectively.**

- For calories, it might be 300. If you choose a time-based goal, you might select 45 minutes. A distance-based goal might be two miles. Remember, you should see different options based on your activity. For example, if you're running indoors, the watch uses the accelerometer, but cycling outdoors uses the GPS on your iPhone to calculate distance. Make sense?

- Depending on the Workout you're in, you may also see Set Pace Alert at the bottom of the screen. If you tap this and start your Workout, Apple Watch alerts you when you're ahead or behind a set pace after one mile.

6. **Tap Open to start the Workout; alternatively, instead of Step 4, just tap the Workout you want, such as Outdoor Cycle, and you'll see a countdown (3-2-1) to start the Workout you've chosen.**

Do your thing and the watch counts your every move. Well, *almost* every move. Remember, Apple Watch might always give you proper credit for things like push-ups, pull-ups, and crunches. Sure, it adds to your Move tab within the Activity app, but it might not help properly calculate your calories burned in the Workout app. But still do them because you know it helps your health — even if your watch doesn't!

See Figure 8-8 for a look at each screen.

7. **Look at your screen for real-time info.**

Your Apple Watch screen shows relevant and real-time information on your Workout (as pictured in the left image of Figure 8-9). With Indoor Walking, for example, it shows you elapsed time, estimated calories burned, speed of average mile, heart rate (beats per minute), and elevation info.

During your Workout, you should see progress updates to help motivate you. You should also receive timely encouragement when you've hit halfway toward the end of your workout, for example, or perhaps based on a milestone, such as reaching one mile during a three-mile jog. See Figure 8-9 for an example.

8. **If you need to pause or end your workout, swipe to the right on the screen.**

You'll see options for Pause, Stop, Lock (to prevent you accidentally hitting your screen during a workout, which could pause or stop the analysis), and New (to start a new session). Tap the relevant button, all shown in the right image of Figure 8-9.

FIGURE 8-8: Workout numbers start at zero so you can select goals. Or don't set a goal and go for a run; Apple Watch still calculates your steps, time, distance, and calories.

9. **To play music while exercising, swipe to the right in the Workout screen and press Play, or skip forward and back between tracks.**

Apple Watch remembers how it played music the last time, such as using Apple Music, your iTunes library, and so on (see Figure 8-10). See Chapter 9 for more on this.

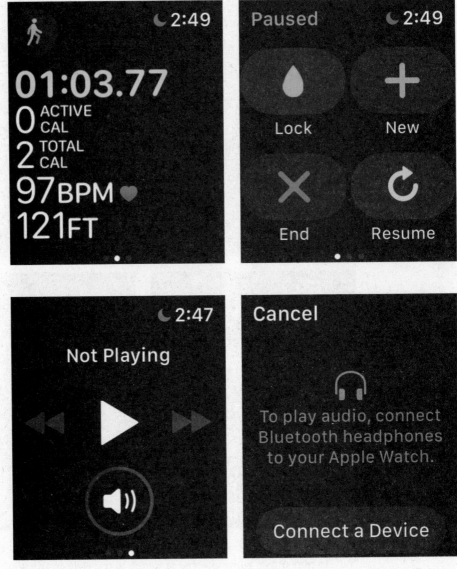

FIGURE 8-9:
Receiving an alert upon achieving a goal (left) can give you incentive to keep going. You can always press the screen to pause or cancel a workout (right).

FIGURE 8-10:
Access your music while working out with a simple swipe to left (left). You may need to tell the watch where to find music to play, such as a streaming service or locally stored tunes (right).

10. **At the end of your workout, view a summary.**

After you press Stop, you'll see a summary of your workout, including total calories burned, active calories burned (when you were physically exerting yourself), resting calories burned, average pace per mile, average heart rate, total distance, total time of workout, and more. Figure 8-11 shows a few of these summary screens after a short walk.

To remind you, all the numbers are color-coded too, such as your total distance in blue, total time in yellow, and active calories burned in pink. At the bottom of the summary screen, you can choose to Save or Discard this information. Tap which option you prefer.

FIGURE 8-11: Swipe for a report on your workout session (left), including distance, calories burned, and heart rate. You're notified and congratulated if you hit your goal (right).

Now you know how to select, start, and stop a Workout, as well as read your summary information. And don't forget: After you review your accomplishments, you can repeat the exercise (with the same goals) to see how you fare, or you might decide to increase or decrease the goal. Or choose a different goal altogether. If you started with time or distance, you might change it up to set a goal based on caloric burn.

Personalizing Reminders, Feedback, and Achievements

Despite what you've heard, information is bliss — not ignorance.

Apple Watch can not only calculate your workouts for you, but it can also present the data in an accessible way, so you can see how well (or poorly) you're doing toward reaching your fitness goals. Actually, the watch goes one step further: It can nudge you to be more active, provide weekly goal summaries, and reward you for a job well done.

Reminders

Apple Watch delivers customizable coaching reminders that can help you reach your Activity goals: Move, Exercise, and Stand. You can disable these in the Apple Watch app on the iPhone. Take a look at Figure 8-12: Go to My Watch ⇨ Activity (left). You can also see a quick-glance summary screen in the Apple Watch app on iPhone (middle), as well as deeper fitness information in the Health app on iPhone (right).

FIGURE 8-12: Get "nudged" into activity by setting reminders to shake your booty (left). View many times you've hit your Activity goals per month, in each category (middle), or get in-depth summaries (right) in the Health app.

Along with notifying you to get up when you've been idle for too long, over time, Apple Watch learns your goals and accomplishments and suggests a daily Move goal for the week that's achievable. You can adjust your fitness goals — whether it's bumping them up or trimming them down — to something more reasonable based on your capabilities or time.

Summary

Every Monday, you should receive a weekly check-in, which serves as a summary of your Activity progress. It might say something like this: "Last week's Active Calorie burn goal was 300. You hit it 4 out of 7 days." On the graph that accompanies the text, you can see which days you reached your goal and by how much; the higher the vertical column, the better you did. You should see all seven days of the week presented.

You can also open the Activity app to glimpse how you're doing per day, and you can always swipe down from within the Activity app to see an hour-by-hour account of your daily Move, Exercise, and Stand goals. See the "Getting Up and Running with the Activity App" section as well as the "Using the Activity App on Your iPhone" section for more on the Activity app.

Achievements

Isn't it enough incentive to know you're doing well? Well, not always. Let's face it: It's nice to be acknowledged for your efforts — and even rewarded.

"Get a pat on the back, right on the wrist," says Apple on its website.

I like that.

Earn special badges, such as the ones shown in Figure 8-13 — although they're a little hard to see because I haven't completed them yet and thus they're not colored in. These badges are stored in the Activity app on your iPhone, which you can look at with pride.

Some examples of Move badges include:

>> **Perfect Month:** Earn this award when you reach your Move goal every day of a single month — from the month's first day to the last.

>> **Move Goal 200%:** Earn this award every time you double your daily Move goal.

>> **100 Move Goals:** Earn this award when you reach your daily Move goal 100 times.

More importantly, perhaps it'll encourage you to keep going.

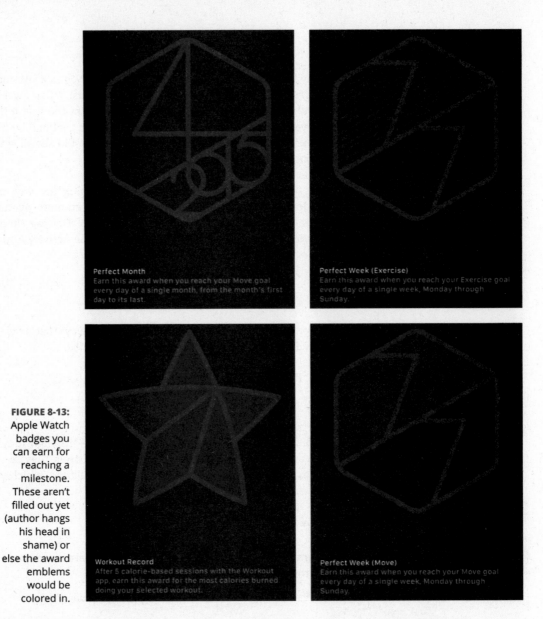

Perfect Month
Earn this award when you reach your Move goal every day of a single month, from the month's first day to its last.

Perfect Week (Exercise)
Earn this award when you reach your Exercise goal every day of a single week, Monday through Sunday.

Workout Record
After 5 calorie-based sessions with the Workout app, earn this award for the most calories burned doing your selected workout.

Perfect Week (Move)
Earn this award when you reach your Move goal every day of a single week, Monday through Sunday.

FIGURE 8-13:
Apple Watch badges you can earn for reaching a milestone. These aren't filled out yet (author hangs his head in shame) or else the award emblems would be colored in.

Using the Activity App on Your iPhone

You can only fit so much information on the small Apple Watch. Thus, the Activity app on your iPhone takes it a step further by providing you with a ton of data based on your Move, Exercise, and Stand achievements as well as your Workout information.

Specifically, the iPhone app's History tab maps your progress over long periods of time; you can look at how well you did by day, week, month, or year. See Figure 8-14 for a look at the Activity app on iPhone.

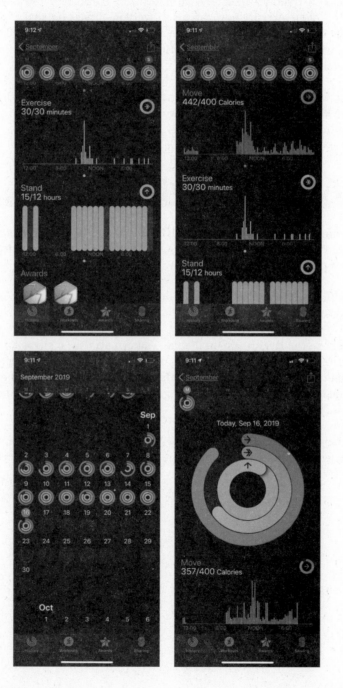

FIGURE 8-14: A look at the Activity app on your iPhone with a weekly summary of your Activity levels.

The Activity app also displays your badges (of honor!) as well as an Achievements tab that shows you various objectives to unlock as added incentives. Like many video games that reward you for achieving goals — such as "Complete an end-level boss fight in under five minutes" or "Capture a town without losing a life" — your Activity app can also give you fitness-related challenges to take on. By default, the app has a dozen challenges, each represented by a badge-like icon, which could be "Reach your Move, Exercise, and Stand goals in one day" or "Exercise a total of 20 miles."

What's more, the Activity app syncs its data with your iPhone's Health app, where it can be accessed by third-party health and fitness apps — with your permission of course.

Checking Activity Trends

Beginning with watchOS 6 in the fall of 2019, Apple Watch users can take advantage of the Trends feature, which gives you a deeper look at your activity (or inactivity), over a period of time, which is viewable on your iPhone.

Specifically, you can anlyze and compare your Activity rings — Move, Stand, and Exercise —to past performances.

Viewed in the Health app on iPhone, you can also see a snapshot of other fitness metrics, including average walking pace, flights of stairs climbed, VO2 max levels (maximal oxygen uptake) and whether you're improving or falling behind. Trends compares your latest 90-day averages to longer-term performance over the past 365 days.

As shown in Figure 8-15, it's all laid out in convient table and graph form, and you even get customized coaching suggestions to get you back on track!

FIGURE 8-15:
A look at the
Activity Trends
data gleaned
from Apple
Watch and sent
to iPhone's
Health app.

Learning to Use (and Love) the Breathe App

 Life can be hectic. And so Apple's Breathe app reminds you to take a moment out of your day to relax and, well, breathe.

Of course, you're always breathing (or else you wouldn't be alive to read this book, silly!), but this wrist-based reminder guides you through a series of deep breaths to calm you down. The Breathe app lets you choose how often and how long you want to breathe, and then let the Apple Watch animation and gentle taps help you focus throughout this short exercise.

Starting a Breathe session

Had a rough day at the office? A crying baby on an airplane? Here's how to initiate a breathing session to calm you down, via your Apple Watch.

1. **Press the Digital Crown to go to the Home screen, and then tap the Breathe app to open it.**

2. **Turn the Digital Crown to set the session's length.**

   Tap Start when you're ready, but be sure to remain still while you breathe. See Figure 8-16.

3. **Inhale as the animation grows on the Apple Watch screen.**

   You also feel little taps on your wrist. Exhale as the animation shrinks and the taps stop.

4. **Breathe until the session ends and your watch taps you twice and chimes (unless Silent mode is on).**

When you're done, you can see your resting heart rate.

FIGURE 8-16:
A look at the Breathe app on Apple Watch. Follow along with the animation and haptic feedback.

 Because you're supposed to be calm and focus on your breathing, your Apple Watch mutes some notifications. If you answer a call or text, or move around too much during a session, the session ends automatically, and you won't get any credit for your efforts.

REMEMBER

Using the Apple Watch app to make Breathe changes

The Apple Watch companion app on iPhone can be used for additional Breathe settings (see Figure 8-17).

For example, when you turn the Digital Crown on Apple Watch to set the session's duration, such as 90 seconds, but you can adjust the default setting from your iPhone's Apple Watch app (under Breathe), as follows:

1. **Open the Apple Watch app on your iPhone.**

2. **Tap the My Watch tab.**

3. **Tap Breathe.**

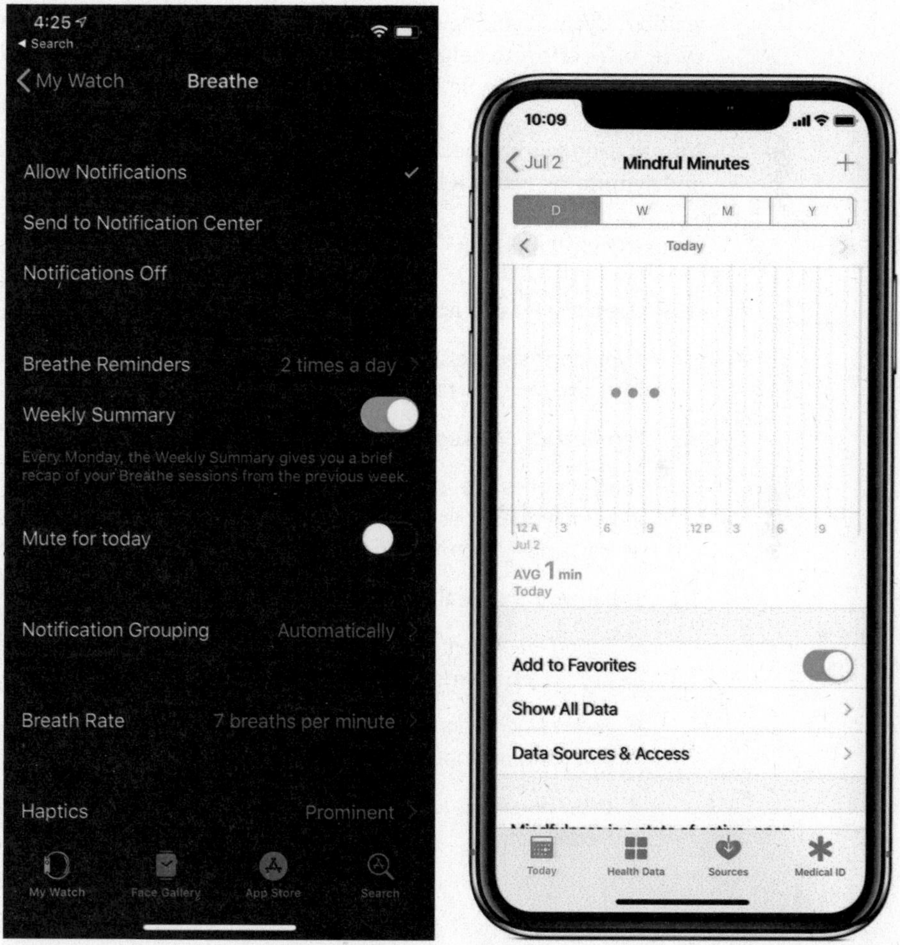

FIGURE 8-17: The Apple Watch app on iPhone is where you make changes to the Breathe app (left); the Health app on iPhone shows a summary of your breathing exercises (right).

4. **Scroll down and tap Use Previous Duration (it's at the bottom of the screen).**

   You can also change the length of each breath — by default it's 7 breaths per minute — and you can tweak how many Breathe reminders you want per day (such as two), how forceful the haptic (vibration) feedback is, and more.

5. **Check your progress.**

   To check how often you take time to use the Breathe app on Apple Watch, track your sessions with the Health app on iPhone. Open the app and select the Health Data tab. Tap Mindfulness, then tap the graph to see more.

Cycle Tracking

If you're female, you'll find one of the newest Apple Watch features at the time of writing, Cycle Tracking helpful. You can gain insight into your monthly menstrual cycle, in an effort to help provide a clearer picture of your overall health. By logging your cycles, your Apple Watch can predict your next period and fertile window.

Information you can access as easily on your wrist can also help track irregularities and symptoms you might want to discuss with your physician.

To use Cycle Tracking:

1. **Open the Health app on your iPhone to set up Cycle Tracking.**

   If you don't see the section, tap Browse in the bottom right of your iPhone and then tap Cycle Tracking.

2. **Tap Getting Started and then Next to answer questions.**

   You'll get questions such as "When did your last period start?" (select date), "how long does your period usually last?" (e.g. 5 days), and "How long is your typical cycle?" (such as 28 days).

3. **Still inside the Health app on iPhone, toggle these options.**

   - *Period Prediction*: Allows the Health app to use the data you entered to predict your period.

   - *Period Notifications:* The Health app will notify you about upcoming periods and send prompts to log.

   - *Other options*: Such as Fertile Window Prediction, Log Fertility, and Log Sexual Activity.

After you've completed the questions and options in the Health app on iPhone, you can now use the Cycle Tracking app on Apple Watch. Open the Health app to log daily information about your menstrual cycle. You can add flow information (Had Flow, No Flow, or Flow Level – Light, Medium, Heavy); record Symptoms (such as Abdominal Cramps, Acne, Appetite Changes, Bloating, Headaches); and glance at cycle length and variation. See Figure 8-18.

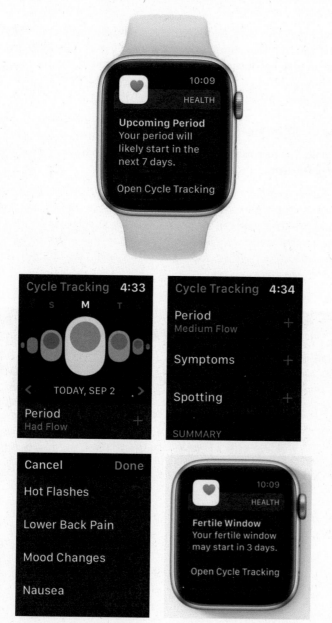

FIGURE 8-18: A look at the new Cycle Tracking app for Apple Watch.

You can even track results from an ovulation prediction kit and readings from a basal body thermometer. All this info will be viewable in a colorful graphical chart in the Health app on your iPhone.

Hearing Health: Using the Noise app

Enjoy hard rock concerts? Work with heavy machinery? Apple Watch's Noise app can show you how loud ambient audio is near you — measured in decibels — and how long you've been exposed to it.

Using the Noise app is quite straightforward:

1. **Tap the Noise app.**

2. **Enable the microphone, which gives the app permission to listen to the environment.**

   It doesn't record or save the audio. The app shows you the estimated noise level around you, measured in decibels (the higher the number, the more damaging the sound can be on your hearing), as shown in Figure 8-19. The number is also complemented with a color-coded system for easy convenience. If it's green, you're good. If it's yellow, turn it down. You can tap to read more info, too.

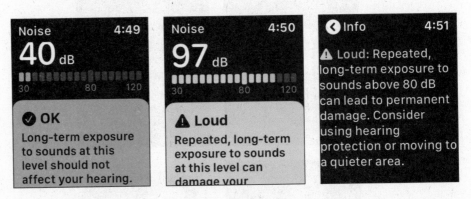

FIGURE 8-19: As Quiet Riot once famously sang, "Cum on feel the noise!" It's easy with the Noise app on Apple Watch.

You can also add the Noise app as a complication to your favorite watch face. See Chapter 4 to find out how.

Advanced Health Help: Heart Rate, ECG, Fall Detection, and SOS

Apple Watch can do a lot more than just provide activity information and workout tracking. Although you now know it measures your resting heart rate and how hard your heart works while exercising, you may not know that it can alert you if something seems a little off. You also might not know that beginning with Apple Watch Series 4, your wrist-mounted device has an electrocardiogram/ECG feature ("EKG," depending on the source), Fall Detection, and Emergency SOS safety features.

Before I get into each feature, I'm sure some of the more technology minded will want to know how Apple Watch does all this! So this section starts with an explanation of each monitor in all its glory.

How the heart rate sensor works

Underneath the Apple Watch case — the part that touches your skin — is a ceramic cover with sapphire lenses. Beneath that is a series of small Taptic Engine sensors, which give you a slight vibration to tell you something, such as when you have a message from someone. In that same area, you also have a heart rate monitor that logs the intensity of your workout, or simply your resting heart rate.

TIP

If someone you know also has an Apple Watch, you can send him or her your heartbeat. See Chapter 5 for more about how to do this, as well as sending taps and sketches to other Apple Watch owners.

The heart rate sensor uses infrared (IR), visible-light LEDs (light-emitting diodes), and photodiodes to detect your heart rate — measured in beats per minute (BPM). The average heart rate is 72 BPM, but when you start your workout and your muscles need more oxygen, your heart beats faster to pump oxygen-filled blood throughout your body.

Every ten minutes, Apple Watch measures your heart rate and stores that information in the iPhone's Health app. This data, along with other information it collects about your movement, goes toward estimating the number of calories you're burning.

The Apple website explains how the heart rate sensor uses "photoplethysmography." It states: "This technology, while difficult to pronounce, is based on a very simple fact: Blood is red because it reflects red light and absorbs green light." Apple Watch uses green LED lights, paired with light-sensitive photodiodes, to detect the amount of blood flowing through your wrist. "When your heart beats, the blood flow in your wrist — and the green light absorption — is greater," says Apple. You can see a map of these amazing sensors in Figure 8-20.

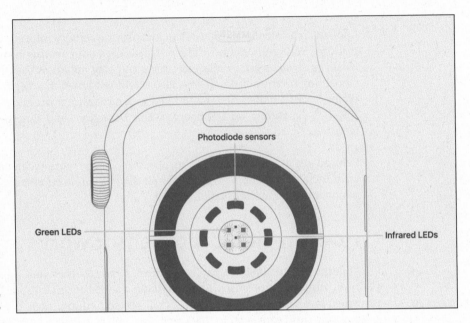

By flashing its LED lights — in an alternating fashion and hundreds of times per second — Apple Watch calculates the number of times the heart beats per minute. Apple says its heart rate sensor can also use infrared light, which it uses to measure your heart rate every ten minutes. But if this infrared system isn't giving a reliable reading, your smartwatch switches to the green LEDs.

For the heart rate monitor to work effectively, you must wear Apple Watch to fit snugly; if your watch band is too loose, the back of the watch case might not touch your skin enough or might move around during a reading. All available Apple Watch bands allow you to tighten things up a bit.

Other factors can also affect a reading.

» **Weather:** Readings might be off if it's too cold out.

» **Movement:** Fast or constant movement, such as an intense game of squash, can jostle the watch around too much for an accurate heart rate measurement.

» **You:** Some people just don't give a good reading at all. Some Apple Watch owners with major tattoos report the heart rate monitor won't work. Why, you ask? The sensor works by shining green light into the wearer's wrist, but can be confused by the presence of dark tattoos inked under the skin.

If you're not getting a good reading, don't forget that Apple Watch and iPhone can work with external heart rate monitors, such as a chest strap, using wireless Bluetooth technology.

Here's a look at it all.

How the ECG monitor works

Apple unveiled its super high-tech electrocardiogram (ECG) feature in the fall of 2018, along with its introduction of the Apple Watch Series 4 models (which you'll need to have to take advantage of this feature). It's also available in Apple Watch Series 5 (2019) and presumably, going forward in all models!

As you might expect from this feature's name, you can now use Apple Watch to detect a dangerously high or low heart rate, or irregular heart rhythm (arrhythmias).

How does it work, you ask? As you know, the back of Apple Watch already measures your heart rate on your wrist. To read ECG pulses, you touch the Digital Crown with the finger on your opposite hand, which, when coupled with pulses captured by the sensors under your watch, creates a closed circuit with your heart. Your watch can now record ECG information and detect when something is off. Figure 8-21 illustrates what happens.

FIGURE 8-21: While it seems like science-fiction, Apple Watch can not only detect and display your heart rate (left), but also becomes an ECG that senses abnormalities (right).

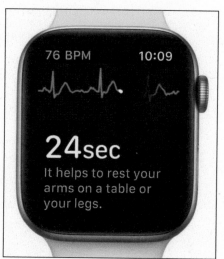

Setting up or editing your Medical ID information

Before you delve in to the various Apple Watch health features, a discussion of how to set up your Medical ID information is in order. Why? Because a lot of these features require this information in the event that something goes wrong with you and you want to inform a close contact, or even 9-1-1, of your emergency.

With that in mind, you need to teach your Apple Watch who is important to contact. You must turn on Wrist Detection for your watch to automatically call emergency services.

Bear in mind falls are also automatically recorded in the Health app on your iPhone, unless you reply that you didn't fall when your Apple Watch asks. To check your fall history, open the Health app on your iPhone, tap the Health Data tab, and then tap Results.

To set up your Medical ID:

1. **Open the Settings app on your Apple Watch, tap General, then make sure that Wrist Detection is enabled.**

2. **Open the Apple Watch app on your iPhone, tap My Watch, and then Emergency SOS.**

   Tapping Edit takes you to the Health app on the iPhone; or, simply open the Health app on your iPhone, then tap the Medical ID tab.

3. **Tap Edit and enter your date of birth and other health info.**

   Depending on where you live, you might also be prompted with questions about Organ Donation (and can sign up here, if you like). See Figure 8-22.

To add emergency contacts, follow these steps:

1. **Open the Apple Watch app on your iPhone, tap My Watch, and then Emergency SOS.**

   Tap Edit at the top of the screen to select various people.

2. **Tap the green plus sign under emergency contacts.**

3. **Tap Contact, followed by their relationship to you.**

   Figure 8-23 shows you the Emergency SOS screen.

FIGURE 8-22:
Use the Health app on iPhone to create a Medical ID that may be sent to emergency responders via your Apple Watch (left). You might also be prompted with other options, like donating your organs (right).

4. **Similarly, tap the red minus button next to the contact to remove an emergency contact, followed by Delete.**

5. **To make your Medical ID available from the Lock screen, turn on Show When Locked.**

   In an emergency, this gives information to people who want to help. Tap Done.

Checking heart rate

At any time, you can check your heart rate using the Heart Rate app. Press the Digital Crown and then open the app and wait a few seconds for Apple Watch to measure your heart rate.

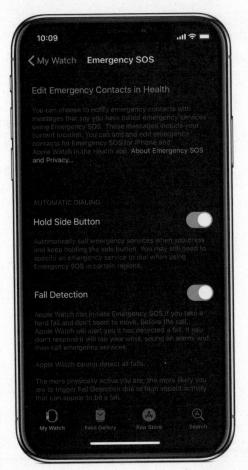

FIGURE 8-23:
You should first set up who your emergency contacts are, just in case you are unable to after falling.

From this screen, you can also view your resting, walking, breathe, workout, and recovery rates throughout the day. Figure 8-24 shows all this in action.

TIP

Alternatively, you can summon Siri to show you heart rate info and/or add the Heart Rate complication to your watch face.

Receiving heart rate notifications

Now here's something you might not know about Apple Watch: If your heart rate remains above or below a chosen BPM number, your Apple Watch can notify you.

FIGURE 8-24: View heart rate info just after you take a reading (left). Or look back at different points of the day through the Apple Watch app (middle). The Health app gives a deeper dive into BPM data (right).

To get going:

1. **On your iPhone, open the Apple Watch app.**

2. **Tap the My Watch tab, then select Heart Rate.**

3. **Tap High Heart Rate, then choose a BPM.**

   If you're unsure what number this should be, consult your physician.

4. **Tap Low Heart Rate, then choose a BPM.**

Now, when Apple Watch measures your heart rate while you appear to have been inactive for a period of 10 minutes, it notifies you if your BPM is higher or lower than your chosen numbers that you set in Step 3 and 4 (see Figure 8-25).

REMEMBER

When you use the Workout app, Apple Watch measures your heart rate continuously during the workout and for three minutes after the workout ends to determine a workout *recovery* rate.

Turning on Fall Detection

Beginning with Apple Watch Series 4 (and any watch after that), Apple added a new and improved accelerometer and gyroscope (another motion sensor that can detect if you've fallen).

FIGURE 8-25:
With the Heart Rate feature activated, you are alerted if your heart rate (BPM) is too low or too high based on parameters you set up in advance.

Do you like the idea of this feature? Can you see a loved one benefiting from it? Here's how to turn it on (or off):

1. **Open the Apple Watch app on your iPhone, and then tap the My Watch tab.**

2. **Tap where it says Emergency SOS.**

3. **Now turn Fall Detection on or off.**

The next section walks you through what to do if you actually fall.

What to do after a fall

When a fall incident occurs, the watch delivers a hard fall alert, taps you on the wrist, sounds an alarm, and displays an alert.

REMEMBER

Although handy, Fall Detection isn't perfect. Apple Watch cannot always detect a fall if you went down slowly. Conversely, it might misinterpret a fast drop as a fall when, in fact, you did something physical.

As shown in Figure 8-26, when Fall Detection is activated, you've got a choice to make:

» **You fell, but you're OK:** You can dismiss the alert by tapping "I fell, but I'm OK."

» **It was a false call:** If you didn't fall at all, tap "I did not fall."

FIGURE 8-26: You'll get some options to tap through if you fall. The example on the right shows what happens if you swipe Emergency SOS. You can call 9-1-1 or someone else.

REMEMBER

>> **If you're not OK:** If you've just had a bad fall, you can tap and drag the Emergency SOS icon. From your watch, this lets you then tap for 9-1-1, or lets you opt for a caregiver, family member, neighbor, friend, or whomever you like.

You must set up your emergency contact list ahead of time as discussed in the section "Setting up or editing your Medical ID information" earlier in the chapter.

>> **If you're unresponsive after 60 seconds:** Apple Watch begins a 15-second countdown while tapping you on the wrist and sounding a loud alert (which gets louder so you or someone nearby can hear it). When the countdown ends, your Apple Watch automatically contacts emergency services. At the same time, it also sends a message to your emergency contacts with your geographical location, letting them know that Apple Watch detected a hard fall and dialed emergency services.

If you call emergency services, you can end the call when your call is finished or, if you no longer need emergency care, by tapping the red Phone icon. On the End Call screen, choose Yes.

REMEMBER

If you've entered your age when you first set up your Apple Watch (or in the Health app on iPhone) and you've indicated you're age 65 or older, the Fall Detection feature automatically turns on.

Emergency SOS

Whether it senses a troubling anomaly through the heart-rate or ECG sensors, or detects a hard fall, Apple Watch can dial 9-1-1, notify your emergency contacts, send your current location, and even display your Medical ID badge for emergency personnel. Or you can initiate a call through your watch if you require some emergency attention.

Apple Watch now lets you make emergency SOS calls internationally. According to Apple, it works "almost anywhere in the world" with cellular versions of Apple Watch Series 5 (and later).

WARNING

For all these scenarios, to make an emergency call, you must have an Apple Watch GPS + Cellular model (with active cellular plan), or be connected to an iPhone via Bluetooth, or have access to Wi-Fi (with Wi-Fi Calling setup, as discussed in Chapter 5).

If you start the countdown by accident, simply let go of the side button. Or if you start an emergency call by accident, firmly press the display, then tap End Call. You may get a call back from an emergency response operator and you can tell him or her you dialed the number by mistake.

To manually call emergency services:

1. **Press and hold the side button on your watch until the Emergency SOS slider appears.**

   It may take two seconds or so.

2. **Continue to hold down the side button.**

   A short countdown to begin and an alert sounds. Alternatively, you can also drag the Emergency SOS slider to the left.

3. **When the countdown ends, your watch automatically calls emergency services, which is usually 9-1-1 for most countries.**

4. **After the call ends, your Apple Watch sends your emergency contacts a text message with your current location, unless you choose to cancel.**

   If Location Services is off, it temporarilies turn on, says Apple.

TIP

If your location changes — such as a first responder taking you to the hospital — your contacts will get an update, and you, too, will get a notification a few minutes later.

To stop sharing your location, tap Stop Sharing in the notification.

Just press and hold the side button and you'll be connected with emergency services.

» **Using Siri to play your music**

» **Pairing Bluetooth headphones with Apple Watch**

» **Listening to music on your Apple Watch**

» **Playing podcasts, audiobooks, and radio plays**

» **Using Apple Watch to control Apple TV**

» **Accessing yourApple Music or iTunes library with your watch**

Chapter **9**

Mucho Media: Managing Your Music, Movies, and More

Music lovers, listen up. Your Apple Watch can help you get more from your favorite tunes. In this chapter, I cover how to use Apple Watch for listening to music while you're on the go. Although I share how to use your watch to control tracks stored on a nearby iPhone, you don't even need your iPhone to listen to music: Your Apple Watch can stream music from the Apple Music streaming service, as well as store and play music on the watch itself.

For the latter two scenarios, there are times when you don't want to bring your iPhone with you, but as long as you have Bluetooth headphones (shown in Figure 9-1) to listen with, you can still enjoy your music.

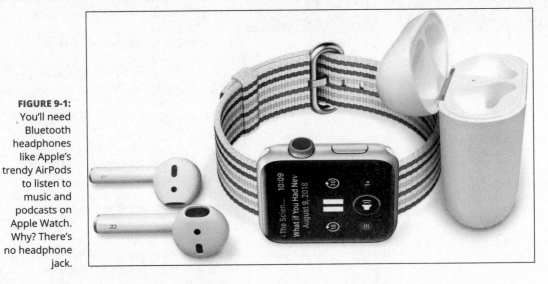

I also discuss how to control your music hands-free — as well as audiobooks, podcasts, and radio plays — by using Siri.

Although not widely publicized, Apple Watch can also control the Apple TV connected to your TV or your Apple Music app (formerly iTunes) on your Mac or iTunes on your Windows PC.

Using Apple Watch to Control Songs Stored on an iPhone

Similar to the iPhone app, Apple Watch has a Music app that lets you find and play music stored on your iPhone.

Sure, you can also install music on the Apple Watch itself — something I cover later in this chapter in the "Playing Music from Your Apple Watch" section. You can also stream music directly from the Apple Music service (more on this later), but most wearers will likely use their watch as a kind of wireless remote control.

Sample scenario: You're walking down the street with your swank Beats headphones or wireless AirPods. Rather than taking your iPhone out of your jeans every time you want to switch tracks or choose a playlist, simply lift your wrist and perform those functions from the comfort of your Apple Watch. After all, the watch is all about convenient, glanceable information when and where you need it. Controlling your music is no different.

To manage your iPhone music from your Apple Watch, follow these steps:

1. **Press the Digital Crown button to go to the Home screen.**

   Here you see all the icons of installed Apple Watch apps.

2. **Tap the Music app.**

   You can also raise your wrist and say, "Hey Siri, play Music" or something more specific. Or if you set up Siri to activate the moment you raise your wrist (see Chapter 7), then you don't need to summon your personal assistant with your voice at all.

   On this Music screen, you see the name of the song, the artist, and the album artwork (if available) as shown in Figure 9-2. If a song is already playing on your iPhone, you see small equalizer bars dance at the top of the screen and some scrolling information, such as the complete name of the track, the album name, and more.

FIGURE 9-2: Swipe up inside the Music app to scroll through album artwork and tap to select that specific track.

3. **Tap the artwork to play the song if it isn't already playing.**

   You hear your selection through your iPhone, wired earbuds/headphones, or Bluetooth headset.

Remember, you can't hear music through Apple Watch — nor would you want to — unless you have a Bluetooth headset or headphones. See the "Pairing a Bluetooth Device with Apple Watch" section for how to add one to your Apple Watch.

4. **Tap + or – to increase or decrease the volume, respectively.**

   A red horizontal bar visually shows you how loud the volume is getting. You can also skip forward or backward between tracks using the double arrows on each side of the Play button. The elapsed time of the track is in the top-left corner and a digital clock is in the top-right corner.

But what do you do if you want to hear a new song? You've got many options on how to handle that.

>> Swiping up inside the Music app displays the album art of other songs in your library.

>> Scrolling all the way to the top opens a menu with the options Shuffle All, Library, and On Phone (to see music content on your iPhone).

>> Tapping the phone or library (which has the music stored on Apple Watch or the Apple Music service) reveals four options: Artists, Albums, Songs, and Playlists; these are discussed in detail in the remaining parts of this section. Another screen is the default that you're currently in, called Now Playing.

Now Playing

This shows you the artist that's currently playing. You can tap the screen to pause the song, or to perhaps skip back to play the song again from the beginning. If no album art has been imported for this track, as recognized by iTunes, this area will just have text and the virtual control buttons, as shown in Figure 9-3.

Be aware there's also a dedicated Now Playing app on your Apple Watch Home screen, which lets you continue playing content on your Apple Watch or iPhone.

Artists

You can see your music collection listed alphabetically by artist/band. Swipe up or down with your fingertip or twist the Digital Crown button at the speed of your choosing to review the list. As shown in Figure 9-4, under each artist, you should see how many songs you have from that person or band.

FIGURE 9-3:
No album art?
No worries.
You should still
see pertinent
information
about a song.

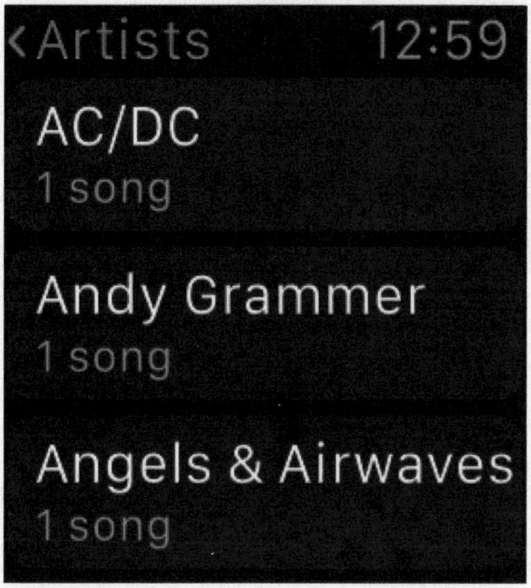

FIGURE 9-4:
The Artists
view inside the
Music app on
Apple Watch.

Albums

As you might expect, this is where you can see all your music listed alphabetically by album name. This is ideal for when you want to hear an entire album from the same artist, or one or more tracks from an album or perhaps a compilation or movie soundtrack with multiple artists.

Songs

As shown in Figure 9-5, this is where you can find all your songs listed alphabetically, regardless of artist. Swipe up or down using your fingertip or twist the Digital Crown button to find something to listen to. Be aware, choosing to see your music collection by song yields the longest list. No worries: You can scroll up or down using the small bar on the top right side of the screen. The small bar appears when you turn the Digital Crown button.

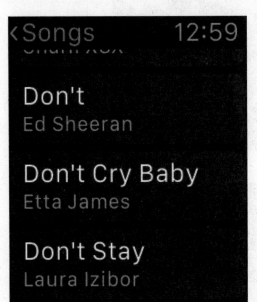

FIGURE 9-5:
Swipe or twist the Digital Crown button to scroll up and down your list of tracks.

Playlists

This shows all your music grouped in some fashion — whether by theme, event, or genre, as shown in Figure 9-6. Windows users can create a music playlist in iTunes, while Mac users have the Apple Music app. Or create your own playlist on your iPhone or iPad. You might call a playlist something like "Driving Tunes," "Workout Mix," "Relaxing Music," or whatever suits your fancy. If you have a playlist on your iOS device, then it's synced to your Apple Watch.

Keep in mind that although you can access a playlist on Apple Watch, you can't create a playlist on it. For that, you have to sync via iTunes or your iPhone.

You can also sync a playlist to the watch to listen to media when no iPhone is around. See the "Playing Music from Your Apple Watch" section for more on syncing playlists.

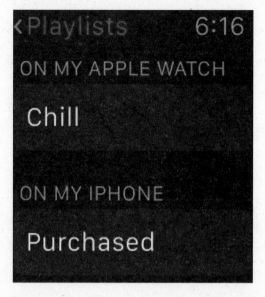

FIGURE 9-6: You can't create a new playlist on your Apple Watch like you can on an iPhone or in iTunes, but you can play it on your watch. You can even store the playlist on the watch if an iPhone isn't nearby.

You can also use Force Touch to display additional options: While listening to a track, press firmly on the screen and you'll see the AirPlay option, which lets you easily stream music to your favorite speakers or TV.

Having Siri Play Your Music

Do you know what's even faster than tapping and scrolling to find music? Asking Siri to play it for you. Say you're itching to hear a song in your collection that's been stuck in your head all day. Or maybe you want to give your favorite band's new greatest hits album a spin from beginning to end? All you need to do is ask Siri (politely) to play an individual song or album. As long as you're in a place you can talk freely, Siri can launch a song, album, artist, playlist, or genre — just by your asking for it. With cellular models, you can do it all without your phone!

To have Siri help you play your music via Apple Watch, raise your wrist and do one of the following three things:

>> Say "Hey Siri, play _____ (name of song, artist, playlist, and so on)."

>> If you enabled Siri to wake up when you lift your wrist (as discussed in Chapter 7), simply raise your wrist and ask to play something.

>> Press and hold the Digital Crown button to activate Siri. Ask away!

After Siri processes your request — and don't forget, you need cellular or Wi-Fi access for this to work — you should see and hear the music you asked for via your iPhone (or, if only stored on Apple Watch, through Bluetooth headphones).

Some examples of what you can ask Siri to play:

» "Play 'Girls Like You.'" (individual song)

» "Play Drake." (artist)

» "Play 'Road Trip' playlist." (specific playlist)

» "Play some hip-hop." (genre)

» "Shuffle my music." (shuffling all tracks)

» "Shuffle The Beatles." (shuffling tracks by a certain artist)

» "What song is this?" or "Who is this?" (Siri shows and/or tells you)

» "Play similar music." (to play similar music to what you're listening to)

You can also control your music with your voice. The following are some commands you can give verbally. Simply press and hold the Digital Crown button (or raise your arm and say "Hey, Siri"), followed by:

» **"Play"** to play the song shown on your watch screen.

» **"Pause"** to pause the track you're listening to.

» **"Skip"** to go to the next track.

» **"Next"** to (also) go to the next track.

» **"Previous song"** or **"Play previous song"** to have Siri play the previously played song.

Pairing a Bluetooth Device with Apple Watch

You can load up your watch with music (see the "Playing Music from Your Apple Watch" section) and take it to go — but there's one catch. You can't listen to music through the Apple Watch's tiny speaker (nor would you want to), and there's no headphone jack to plug in headphones. Instead, you need to pair Bluetooth headphones (or a speaker) to Apple Watch to hear music.

REMEMBER

To clarify, this isn't necessary if your iPhone is nearby because you should hear music coming from the iPhone's speakers or from headphones connected to the iPhone. Pairing a Bluetooth device with Apple Watch directly is only required if your iPhone isn't around.

To connect a Bluetooth device to your Apple Watch, follow these steps:

1. **Put your Bluetooth headphones or speaker in pairing mode.**

   You might see a flashing light on the device to confirm it's awaiting a connection. It's different for all products, but pairing usually involves pressing and holding the Home button until you see a flashing light.

2. **On Apple Watch, locate the Settings app — which looks like a gray gear — and tap it.**

   Alternatively, you can lift your wrist and say "Hey Siri, Settings."

3. **Inside Settings, twist the Digital Crown button until you locate Bluetooth and then tap this option.**

   Wait a moment and your Bluetooth headphones or speakers should appear in this list.

4. **Select the Bluetooth device you want to pair it with by tapping its name with your fingertip.**

   Figure 9-7 shows Apple Watch searching for a device.

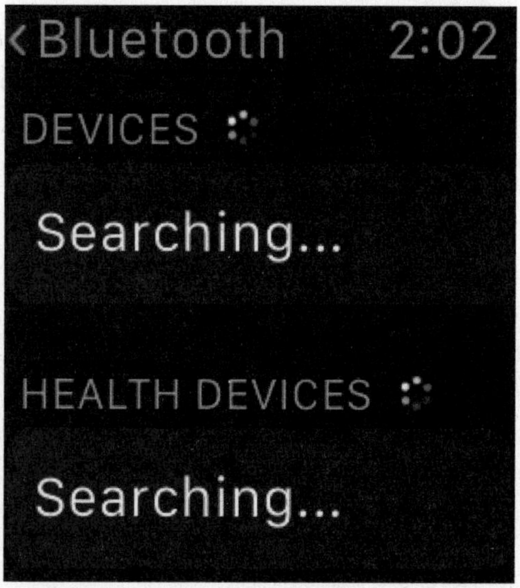

‹Bluetooth 2:02

DEVICES ⋮

Searching...

HEALTH DEVICES ⋮

Searching...

FIGURE 9-7:
Pair a Bluetooth device to hear synced music on your Apple Watch if no iPhone is nearby.

5. **When a device appears, you can tap it.**

You've successfully paired the device! However, you still need to change the music source from iPhone to Apple Watch to hear music come from your Bluetooth-enabled headphones, which you can learn about in the "Playing Music from Your Apple Watch" section.

Streaming Apple Music to Your Apple Watch

As you may or may not know, Apple Music is a monthly subscription service from Apple that lets you listen to more than 45 million songs. Unlike paying for and downloading individual songs or albums from iTunes, Apple Music is a streaming service, so all the songs are stored on Apple's servers in the "cloud" for you to access from your device (as if they were stored locally). Neat, huh? Instead of "owning" the music, you're essentially renting the songs, if you will, with your monthly fee.

In other words, it's not an "a la carte" model, but rather a buffet-style "all-you-can-eat" approach, giving you unlimited access to many songs, with your one monthly fee. To use this feature, you must have an Internet connection. There is an option to download some Apple Music tracks to an iPhone, iPad, Mac, or PC in the event you know you'll be without the Internet for a while.

Another part of Apple Music is an Apple Radio feature, which might introduce you to new songs through curated playlists you can stream on-demand.

Simply open the Music app on your Apple Watch and scroll up, then tap Library. You can also use Siri to play any song in the Apple Music catalog and listen to custom stations.

Playing Music from Your Apple Watch

Not all smartwatches let you store music on your wrist, so you do need a nearby smartphone, but Apple Watch does indeed offer this feature. After all, sometimes, you might not have or want your phone with you, such as when you go out for a jog for a few minutes. If that's the case, why shouldn't you listen to your favorite high-energy music to keep you pumped up?

This convenience comes with two limitations:

>> **Sound:** As stated before, Apple Watch has no headphone jack. You need a pair of Bluetooth headphones wirelessly paired with your smartwatch if you want hear anything stored on it. See the section "Pairing a Bluetooth Device with Apple Watch" for more on this.

>> **Space:** Although you have some storage allocated for many songs, this space is considerably smaller than what your iPhone, iPad, and computer have. Specifically, you've got up to 2 gigabytes of internal storage on Apple Watch dedicated to music and podcasts, which is the equivalent of about 500 songs (at roughly 4 megabytes each).

If you accept these caveats, follow these steps to transfer music to your Apple Watch from an iPhone:

1. **Connect your Apple Watch to your PC or Mac via its USB charger.**

Use the special magnetic charger that shipped with your Apple Watch.

2. **On your iPhone, open the Apple Watch app.**

3. **Under My Watch, scroll down and tap Music.**

You have to scroll down a bit after the Music screen opens because of its many settings, shown in Figure 9-8, along with apps installed on your Apple Watch.

4. **You will see Heavy Rotation already listed under the Automatically Add section.**

This synchronizes the playlists and albums you're listening to with Apple Watch.

5. **To add new albums and playlists, press the orange + icon, next to the words "Add Music."**

6. **Now you can search through music on your iPhone using the search window or by browsing through sections.**

The section will include Artists, Albums, Genres, and Compilations. You should also see custom-made playlists you've already created on your iPhone or iTunes (such as "Awesome Driving Tunes," or "Reggae Mix").

7. **Select what you want to transfer.**

Based on which selection you make by tapping on what music to sync over, the relevant songs are transferred to your Apple Watch. You should see the words Sync Pending and then Synced. Keep in mind that the more songs you synchronize, the longer it takes to copy them to Apple Watch.

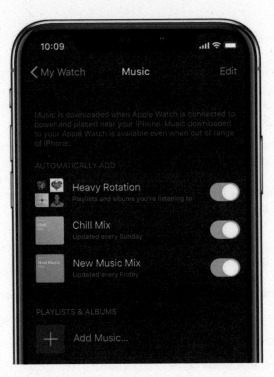

FIGURE 9-8:
Sync music
between your
iPhone and
Apple Watch.

**TECHNICAL
STUFF**

Don't worry about transferring too many tunes over to Apple Watch because Apple won't let you go over your allotted storage of 2 gigabytes (or about 400 to 450 songs).

Have songs stored on your Apple Watch? To listen to music on your Apple Watch rather than your iPhone, follow these steps:

1. **In the Music app, scroll up through your album art and select Library.**

   Here you should see options on where to play your music from.

2. **Tap Device and you should see you're able to toggle between your devices.**

   The screen says: "Choose a music source to play from: iPhone or Apple Watch."

3. **Select Apple Watch, as shown in Figure 9-9.**

   After you tap Apple Watch, the screen should say "Pair a Bluetooth headset to listen to music on your Apple Watch."

FIGURE 9-9:
You need to
select Apple
Watch as your
music source
to hear music
when your
iPhone isn't
around.

4. **Tap Settings and then choose a paired device from the list; pick the one you've got with you.**

Now you're ready to rock, roll, dance, swing, or relax.

Removing Music from Your Apple Watch

If you've grown tired of some songs on your Apple Watch, here's how to remove them:

1. **On your iPhone, open the Apple Watch app, then tap the My Watch tab at the bottom left of the screen.**

2. **Scroll to Music and tap it, then tap Edit in the upper-right corner.**

   You only see Edit if you have music synced to your watch.

3. **Under Playlists & Albums, tap the red minus icon next to any music that you want to remove, then tap Delete (see Figure 9-10).**

   You can also turn off any automatically added playlists that you don't want on your watch from this screen.

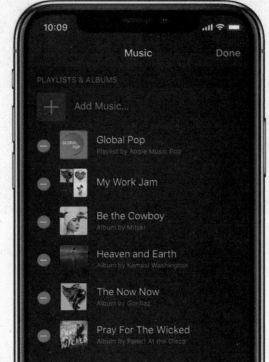

FIGURE 9-10:
It's just as
easy to take
songs off your
Apple Watch
as it is to add
them, but both
require your
iPhone.

Playing Podcasts, Audiobooks, and Radio Plays

Your Apple Watch can also be used to enjoy other kinds of audio entertainment — not just music — including podcasts, audiobooks, and radio plays.

This part of the chapter shows how to pull it all off.

Podcasts

Podcasts (a word that fuses iPod with broadcasting) are free downloadable programs from the Internet. These include comedy routines, news reports, political rants, cooking instructions, tech advice, gardening tips, religious sermons, or the latest music remix from a popular DJ. Many podcasts have video, but in this section, I'm referring only to the audio-based ones.

Unlike your favorite local radio station, podcasts let you to choose when the program starts (called *time shifting*) and where you want to listen to it — even in another country (called *place shifting*). Anyone can publish a podcast — be it a 16-year-old video game fan or huge media corporations, such as ABC/Disney, CNN, or HBO. Thousands of radio stations have podcasts of their popular programs.

Best of all, they're free.

You can download podcasts to your computer (via Apple Podcasts on Mac or iTunes on a Windows PC), onto an iPhone or iPad, and even onto your Apple Watch.

To download a podcast to your iPhone, follow these steps:

1. **Tap the Podcasts icon.**

   You'll see Top Episodes, Top Podcasts, Editor's Choice, and New & Noteworthy.

2. **Click a category to choose a specific theme, or click to play a podcast right in iTunes.**

3. **If you hear a podcast you like, you can subscribe with a simple click of the mouse.**

   Every time a new show is available, you can have it automatically download to your computer's hard drive (or phone's internal memory).

Now it's time to sync podcasts to your Apple Watch!

Syncing podcasts to your Apple Watch

This is really easy stuff, folks. In fact, all your Apple Podcast subscriptions sync to your Apple Watch when it's charging! It may take a few minutes to transfer over a few podcasts, so be patient.

When you want to hear a podcast, tap the Podcast app on your Apple Watch Home screen.

This icon is purple with a white symbol inside, and you can see all your subscribed podcasts. You can twist the Digital Crown if you have many of them, and the images for each one don't all fit on the screen. See Figure 9-11 for an example of what a podcast looks like on Apple Watch.

Alternatively, raise your wrist and ask Siri to open a particular podcast.

FIGURE 9-11:
Apple Watch
makes it super
easy to play
podcasts.
Just tap the
Podcast icon
on the Apple
Watch Home
screen (or
use Siri to
open the app)
to hear your
podcasts.

While inside the Podcast app, you can do the following:

>> **Move among your podcasts:** Flick the screen to navigate between your podcasts.

  If you swipe down, you'll see an option to play podcasts stored on your iPhone or the Library tab, which opens individual podcasts on your Apple Watch (see Figure 9-12).

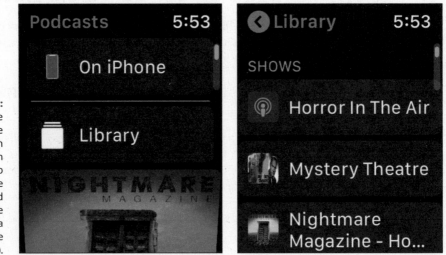

FIGURE 9-12:
Once you're
inside the
Podcast app on
Apple Watch
(left), swipe up
to access the
library stored
on your Apple
Watch or on a
nearby iPhone
(right).

>> **Play a podcast through a compatible speaker or TV:** Press and hold on the screen (Force Touch) and choose AirPlay.

>> **Manually sync your podcast:** Podcast episodes are automatically removed from your Apple Watch after you listen to them, and shows that you subscribe to refresh automatically whenever new episodes are available. If you want to manually choose podcasts to sync, however

 a. *Open the Apple Watch app on your iPhone.*

 b. *Go to My Watch ⇨ Podcasts, and tap Custom.*

 c. *Choose shows with episodes in your Podcasts Library to sync to your Apple Watch.* Figure 9-13 shows what this looks like.

Audiobooks

Call audiobooks today's answer to books on tape — if you're old enough to remember those! As you might expect, audiobooks are usually spoken versions of books read by a narrator as if you were being read a bedtime story. This differs from a radio play, which I discuss in the "Radio plays" section, where a cast acts out a performance, often with sound effects and music.

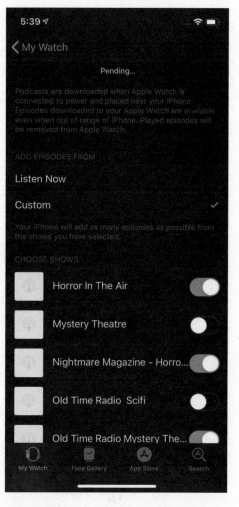

FIGURE 9-13:
If you don't want all podcasts automatically synced to your Apple Watch, go into the Apple Watch app and under the Podcast settings, manually choose what you'd like.

You can download millions of audiobooks from the Internet, but unlike podcasts and radio plays, they're usually not free. Still, they're a great way to have a book read to you — if you're visually impaired, if you want to make a long commute in the car (or on the bus or train) a more enjoyable one, or if you have trouble reading and would simply rather hear someone read to you.

Audiobooks are available for new books (including *New York Times* bestsellers) as well as slightly older and even classic titles. It doesn't matter if it's fiction or nonfiction; chances are, you can find an audiobook version of a paperback or hardcover book and many electronic books (ebooks) too.

There are also audiobook services such as Audible (now owned by Amazon) that support Apple Watch, so you can download and listen to these straight from your wrist-mounted companion. Figure 9-14 shows you the Apple Watch Audible option. You can also find more information about this on Audible.com.

FIGURE 9-14:
The easiest way to play an audiobook on Apple Watch is to use an app such as Apple's own Books (on iPhone) or a service such as Audible.

Okay, here's how to download or stream audiobooks on Apple Watch.

You need to first purchase audiobooks with your iPhone before you can listen to them on your Apple Watch, but beginning with the watchOS 6 operating system, you no longer need to start the audiobook on your iPhone to continue listening on Apple Watch.

Here's the easiest way to play an audiobook on your Apple Watch:

1. **Launch the Books app on your iPhone and tap the Explore Audiobooks tab.**

2. **Scroll to see what's New & Trending, peruse the Top Charts, search by keyword, and more.**

3. **Download an audiobook you want to listen to.**

4. **Tap the Audiobooks icon on Apple Watch.**

   You see all available audiobooks listed here. See Figure 9-15.

FIGURE 9-15:
Compared to a
couple of years
ago, it's now
easier to play
audiobooks on
Apple Watch.

5. **Choose how to listen.**

Inside the app are On iPhone and Library options. Choose Library if you'd like
to play Apple Books audiobooks directly from your Apple Watch.

You have other ways to play audiobooks on your wrist; just remember that copying a downloaded audiobook to a music folder in iTunes makes it easier to control
and play on an Apple Watch, but might be a bit more difficult to do if the audiobook has digital rights management (DRM) encryption on the file. DRM might
limit or prevent the file from copying to another device. (DRM protects the copyright owner to help reduce piracy, which is the unauthorized distribution and/or
duplication of copyrighted material.)

There's another way you can access audiobooks on Apple Watch. Many thousands
of human-read or computer-generated audiobooks based on classic works are
freely available in the public domain that you can copy over to Apple Watch. While
it's a little more work than the aforementioned instructions,you can copy these
freely (and legitimately) available audiobooks to your Apple Watch from such
websites as Loyal Books, LibriVox, and Project Gutenberg.

After you've downloaded the audio files — usually MP3s — to your Windows PC hard drive, follow these steps to sync them to your Apple Watch:

1. **Launch iTunes and click the Music tab.**

   In the top-left corner, you should see a little icon with an arrow.

2. **Click this icon to display some options, including New.**

3. **Click New, followed by New Playlist.**

4. **Name your audiobooks playlist, such as "Audiobooks."**

5. **Wherever you downloaded the DRM-free audiobooks to your computer — maybe your desktop or a Downloads directory — drag and drop them into this newly created playlist.**

   Now you're ready to sync your new playlist to Apple Watch so you can hear your audiobook(s) without a nearby iPhone.

Radio plays

If you're listening to only music on your iPhone, iPad, or Apple Watch, you're missing out on many thousands of downloadable and free dramas and comedies to help keep you entertained while on the go.

Popularized in the 1940s before TV took off, radio plays — or old-time radio (OTR) shows, as they're often referred to today — are enjoying a 21st-century revival thanks to the Internet and MP3s. A whole new generation of listeners can now experience these wonderfully written and performed "theater of the mind" episodes.

Recommended shows include the creepy *Inner Sanctum* and *The Price of Fear* (with Vincent Price) mysteries; nail-biting adventures from *Suspense* and *Escape*; the hilarious antics of Jack Benny, Abbott and Costello, and sci-fi classics, such as *Journey Into Space* and *X Minus One* (featuring many Ray Bradbury yarns).

Filling up your digital devices with these timeless tales is as easy as subscribing to one of the many dozen OTR podcasts (some with daily updates) or by bookmarking such websites as Archive.org, RelicRadio.com, and OTRCat.com — each of which has thousands of free downloadable episodes.

Because most of these older shows have copyrights that have long since expired (or didn't have any to begin with), they're now available for free through the public domain. Some are newer, such as many BBC radio plays and recently published radio dramas based on *The Twilight Zone*, and are still protected by copyrights that prohibit you from copying and distributing them. When in doubt, contact the website that houses these audio plays.

To sync radio plays to your Apple Watch, follow these steps:

1. **Launch the Music app and tap the Playlist tab.**

2. **Tap the "+" sign, for New Playlist.**

3. **Give a name to your Radio Shows Playlist, such as "Radio Shows" or "OTR," or by name, such as "Inner Sanctum" or "Suspense."**

4. **Tap "Add Music," followed by Library, and then select the folder that has the radio plays (such as a Download folder).**

   You're ready to sync your new playlist over to Apple Watch so you can hear your radio plays without a nearby iPhone.

Now you can look forward to — rather than dread — your daily commute.

Controlling Apple TV and iTunes with Your Apple Watch

Your fancy schmancy new Apple Watch can control your iPhone wirelessly. It can play music or snap the shutter on your iPhone's camera (see Chapter 12). More importantly, it can also manage the Apple TV box connected to your TV or the iTunes software installed on a Windows PC (on a Mac, it's now called Apple Music).

With the latter, the Remote app lets you play back content on your computer as if it were a TV — as long as you have media in your iTunes library. So, put that mouse away and start controlling iTunes from your wrist! Or maybe you're cooking in the other room and you want to change the music pumping from the speaker? No need to physically go to your laptop or desktop.

Before you can control iTunes or Apple TV, though, you first need to set up Home Sharing in iTunes and sign in with your Apple ID.

Setting up Home Sharing

To set up Home Sharing in iTunes on a Windows PC, follow these steps:

1. **Download the free Remote app from the App Store.**

2. **On your computer, open iTunes and click the little rectangular icon in the very top-left corner of the screen.**

3. **Click Preferences.**

A number of tabs appear across the top of this Preferences window.

4. **Click the blue Sharing tab and select what media you'd like to share via Home Sharing.**

Options include Music, Movies, Home Videos, TV Shows, Podcasts, iTunes U, Books, Purchased, and others.

5. **Connect your devices, such as an iPad and iPhone, to your home Wi-Fi network and then sign in to Home Sharing via the Remote app.**

Home Sharing is now enabled on iTunes.

Controlling Apple TV remotely

To control Apple TV from your Apple Watch, follow these steps:

1. **Press the Digital Crown button to go to the Home screen.**

2. **Tap the Remote app.**

You can also raise your wrist and say "Hey, Siri, Remote." Either action launches the Remote app. You're presented with two options: Apple TV and iTunes.

3. **Select Apple TV if you own one of these media boxes and it's attached to your TV.**

As long as your iPhone and Apple TV box are joined to the same wireless (Wi-Fi) network at home or work, you can access your Apple TV as if your Apple Watch were that small remote control that comes with it. (You can also control Apple TV with the Remote app for iOS — for the iPhone, iPod touch, and iPad — if you'd like to.)

4. **Tap one of the four arrows on the watch — Up, Down, Right, or Left — to navigate Apple TV's menus.**

No more reaching for the hardware remote or phone or tablet; controlling your media can now be done with a tap on your wrist.

5. **Tap the center Play/Pause button to stop and start your media at your convenience.**

You're at home, with your feet up on the coffee table, and you're enjoying a bit of Netflix on your big screen — until your dog brings you a leash in its mouth (and with those sad eyes). Tap the center of your Apple Watch to pause playback and then take your pup for a walk.

6. **Press the Digital Crown button to return to the Home screen.**

And don't forget to turn off your TV and Apple TV box if you're done. Don't waste power and money.

Controlling iTunes remotely

To control iTunes (on a Windows PC) from your Apple Watch, follow these steps:

1. **Tap the Digital Crown button to go to the Home screen.**

2. **Tap the Remote app.**

You can also raise your wrist and say "Hey, Siri, Remote." Either action launches the Remote app.

3. **Inside the app, tap iTunes.**

You're prompted to enter a four-digit code to access your iTunes library. And remember, you need to be connected to the same Wi-Fi network as your Windows PC and be signed into Home Sharing in iTunes (as covered earlier in this section).

4. **Swipe around the Apple Watch screen to access the library of content on iTunes and then select something to play.**

The Remote app lets you control your iTunes library from anywhere in your home. Fast-forward, pause, or skip back a track or two. Choose a song, shuffle an album, or select a custom playlist. And if your Mac or PC is sleeping, opening the Remote app wakes it up.

» **Using Apple Pay on Apple Watch**

» **Setting up Apple Watch for Apple Pay**

» **Paying with Apple Watch without a nearby iPhone**

» **Accessing Wallet on Apple Watch**

» **Using Apple Pay Cash to send and receive money on Apple Watch**

» **Using Apple Watch for deals and rewards**

Chapter **10**

Pay for Play: Making Mobile Payments with Apple Watch

This chapter looks at how to use Apple Pay on Apple Watch, as well as Apple Pay Cash, to send friends money. First, though, a brief primer on what Apple Pay is.

If you own an iPhone 6 or newer, then perhaps you've had a chance to try out Apple Pay (www.apple.com/apple-pay), Apple's proprietary mobile payment solution that lets you buy goods and services without needing your wallet.

Simply wave your compatible phone over a contactless terminal at a participating retailer — with your finger on the Touch ID sensor (built into the Home button) or with newer models, looking at your iPhone's camera using Face ID technology — and the transaction is completed. This saves time at retail because you don't have to dig for the exact change or deal with finding the right credit or debit card in your wallet.

Intrigued? Read on to find out how to set up and use Apple Pay, how to use Apple Pay Cash, how to send cash to friends, and how to access Apple Watch-specific deals and rewards.

Apple Watch and Security

Before I talk about paying for something on Apple Watch, a word on Apple Pay on iPhone.

If you're worried about security, Apple Pay (shown in Figure 10-1) doesn't disclose your financial information to the retailer; it uses a unique numerical *token* rather than actual credit or debit card numbers shared with the store (which could put your info at risk). In fact, Apple claims using Apple Pay is safer than using a physical card.

FIGURE 10-1:
Leave your wallet at home and use the Wallet app instead! Simply tap your iPhone or Apple Watch at retail to buy something securely.

Because of the integrated near field communication (NFC) antenna in the latest iPhones, you don't need to open an app or even wake up your phone to use it. As long as your one-of-a-kind fingerprint or face is detected — two examples of biometrics technology that prove it's you and only you — you should feel a subtle vibration on the iPhone, and hear a small beep to confirm the *digital handshake* has been made.

You can also use Apple Pay to pay for apps in the online App Store. For this, you can also use the Touch ID or Face ID sensor built into iPhones, iPads, and MacBook Pro (but you can't use the tablet or computer to pay at retail). Checking out is as easy as selecting Apple Pay from the list of options and using your finger or face to identify you.

Before you can use Apple Pay — on any device — you first need to establish your Apple Pay account. You'll do this first on iPhone and then Apple Watch.

Setting Up Apple Pay on iPhone

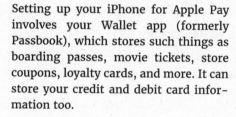

Setting up your iPhone for Apple Pay involves your Wallet app (formerly Passbook), which stores such things as boarding passes, movie tickets, store coupons, loyalty cards, and more. It can store your credit and debit card information too.

I get to setting up Apple Pay on Apple Watch shortly, after covering the other Apple devices first. To set up Apple Pay on your iPhone, follow these steps (oh, and iPad and MacBook instructions are embedded here too):

1. **Tap the Wallet app icon.**

   This launches the Wallet app.

2. **Tap the + sign in the upper-right corner.**

   This lets you add a new card. On a compatible iPad, tap Add Credit or Debit Card. Alternatively, you can also go to Settings, choose Wallet & Apple Pay, and then select Add Credit or Debit Card. On a compatible MacBook Pro with Touch ID, choose System Preferences ⇨ Wallet & Apple Pay, and turn on Allow Payments on Mac. See Figure 10-2.

FIGURE 10-2:
It's a breeze to set up your iPhone with Apple Pay. The step-by-step instructions should take you less than three minutes (seriously).

3. **If you want to use the same credit or debit card that's tied to your iTunes/App Store account, enter the card's security code.**

   As you likely know, it's located on the back of your card.

4. **To add a new credit or debit card, use your device's main (rear-facing) camera to capture the information on your credit and debit card or choose to enter it in manually.**

   You have to fill in additional information it might ask for, such as the security code.

5. **Your bank then verifies your information and decides if you can add your card to Apple Pay.**

   Not all banks are supported, but many are. You may be prompted to provide additional information to verify it's you for your own security and privacy.

6. **Tap Next after your card is verified.**

   You can now start using Apple Pay at supported retailers and in the App Store. Keep in mind that the first card you add is your default payment card, but you can go to the Wallet app anytime to pay with a different card (or select a new default in Settings ⇨ Wallet & Apple Pay). Otherwise, you don't need to do anything. See the "Looking at the Wallet App" section later in this chapter for more on changing your default card and other options.

TECHNICAL STUFF

Whenever you use your credit or debit card at a retailer, your card number and identity, including your name, is visible to the retailer. This can cause privacy and security risks. However, Apple Pay uses a unique Device Account Number (DAN) for each transaction. This DAN is generated, encrypted (securely blocked from outsiders), and then stored in the Secure Element — a dedicated chip inside the iPhone, iPad, and Apple Watch. Apple says these numbers are never stored on Apple servers. The DAN and a dynamically generated security code are used to process your payment when purchasing something; therefore, your actual credit or debit card number is never shared by Apple with the retailer.

What happens if you lose your iPhone, iPad, or Apple Watch? Don't worry about them being used to make a fraudulent purchase. The crook will need your unique finger on the Touch ID sensor (Home button), your face for Face ID, or a PIN for Apple Watch. If you ever misplace your phone or tablet, you can use the Find My iPhone feature on the iCloud website (www.icloud.com) or on another iOS device to quickly put your device in Lost Mode to suspend Apple Pay. Alternatively, you can completely wipe your device clean of any data. And yes, you can now find your Apple Watch using the Find My app on iOS devices (and iCloud.com). So long as your Apple Watch has power, it will show up on the Find My map on an iPhone or iPad.

Setting Up Apple Pay on Apple Watch

As shown in Figure 10-3, the process to set up Apple Pay on Apple Watch is similar to the other Apple products:

1. **Open the Apple Watch app on your iPhone, and tap on the My Watch tab (lower-left corner).**

   If you have multiple watches, choose one (for me it says "Marc's Apple Watch, 44mm").

2. **Tap Wallet & Apple Pay.**

3. **Tap the orange words Add Card, and follow the steps to add a card.**

4. **To add a new card, tap Add Credit or Debit Card.**

   If you're asked to add the card that you use with iTunes, cards on other devices, or cards that you've recently removed, choose them, then enter the card security codes.

FIGURE 10-3: You only need to activate a credit or debit card once on your Apple Watch, then you're good to go.

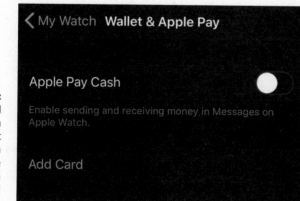

5. **Tap Next.**

   Your bank or card issuer will verify your information and decide if you can use your card with Apple Pay. If your bank or issuer needs more information to verify your card, you'll be prompted for it. When you have the info, go back to Wallet & Apple Pay and tap your card.

6. **After your bank or issuer verifies your card, tap Next.**

   You can now start using Apple Pay (see the next section).

Using Apple Pay with Your Apple Watch

Because you're wearing Apple Watch as opposed to holding it, Apple Pay on Apple Watch is even easier and faster to use at retail than an iPhone. Simply wave your Apple Watch over a contactless terminal at a supporting retailer.

Those wearing an Apple Watch can keep their phones tucked away in a purse or pocket — or even left at home if preferred — but the payment process is similar.

To pay for something with your Apple Watch, follow these steps:

1. **Step up to a contactless reader (point-of-sale terminal).**

   Yes, these are the same ones that support smartphones and NFC-enabled cards, such as Visa's PayWave, MasterCard's PayPass, and Amex's ExpressPay. The person behind you might not be sure what you're doing. Just smile and proceed to the next step.

2. **Double-tap the side button on Apple Watch.**

   You're now ready to make the *digital handshake*.

3. **Hold the face up to the contactless reader, and within a second or two, a tone and slight vibration confirms your payment information has been successfully sent.**

   You won't need to open an Apple Pay app on the watch or anything like that. No wonder Apple calls it "Your wallet. Without a wallet."

4. **If you need to choose Credit or Debit on the terminal, choose Credit.**

   Depending on the store and transaction amount, you might need to sign a receipt or enter your PIN.

You will, though, have to open the Wallet app on Apple Watch if you want to switch cards while paying. To do that, follow these steps:

1. **Double-tap the side button to display Wallet.**

   You can also say "Hey Siri, Wallet" or press the Digital Crown button and find and tap the Wallet app. This is a similar process to paying using Apple Pay, but don't hold your watch up to the contactless reader just yet.

2. **Swipe up or down with your fingertip or twist the Digital Crown button to browse through your cards.**

3. **Select the card you want.**

4. **Hold the Apple Watch face near the NFC reader to pay. See Figure 10-4.**

   You should feel a slight vibration and hear a faint chime to confirm the transaction was successfully completed.

FIGURE 10-4: At a contactless reader, tap your iPhone or Apple Watch to use Apple Pay (left). On Apple Pay for Apple Watch, press the side button twice and tap to pay (right).

REMEMBER

Apple Watch works with any iPhone 5 or newer; iPad Pro, iPad Air, iPad, and iPad mini models with Touch ID or Face ID; and Mac models with Touch ID or any models introduced in 2012 or later with an Apple Pay-enabled iPhone or Apple Watch nearby. You can use Apple Pay with a compatible Mac for payments on the web (with the Safari web browser).

Paying without a Nearby iPhone

Remember, Apple Pay differs from most other mobile payment solutions because your credit or debit card number is never visible to the retailer and is thus safer to use. When you add your card information to Apple Watch, that unique and encrypted Device Account Number (DAN) is assigned and stored on a dedicated chip inside the watch (that Secure Element technology, like with the newer iPhones and iPads).

"But wait a sec," you're thinking. "Apple Pay on iPhone requires my fingerprint on the Touch ID sensor or my face for Face ID (for iPhones that don't have a Home button), so how do I do that on Apple Watch?"

To enable Apple Pay on the watch, you need to create a four-digit passcode using the companion Apple Watch app on your iPhone. This passcode is used to authorize Apple Pay whenever you put the watch on your wrist. Go to My Watch ➪ Passcode to enable and create a passcode using the virtual keyboard, as shown in Figure 10-5.

And as you might guess, those sensors on the back of the watch aren't just used for your heartbeat; they *know* whenever the watch has been taken off. And you must once again type your secret passcode when you put the Apple Watch back on your wrist.

Clever, eh?

That way, if someone puts on your Apple Watch or tries to use it without slapping it on his or her wrist, your information is safe because the other person won't know your passcode.

FIGURE 10-5:
Use your iPhone to create your four-digit passcode for your Apple Watch (left). You are prompted to enter one on a virtual keyboard (right).

Looking at the Wallet App

Just like the app on your iOS devices — iPhone, iPad, or iPod touch — Wallet is available on Apple Watch, and it also lets you store your airline boarding passes (with scannable QR code), movie and sports tickets, coupons, loyalty cards, and much more, as shown in Figure 10-6. Think of it as a handy digital version of your actual wallet (if you still use one).

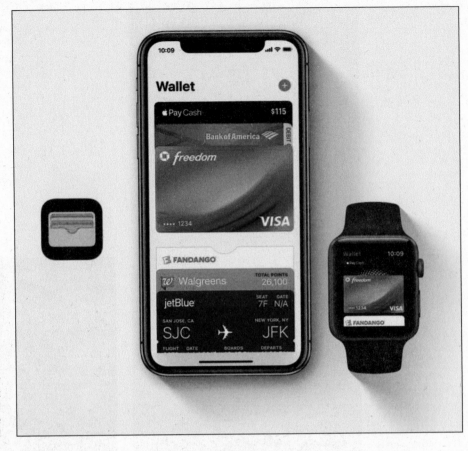

Other sections in this chapter talk about how Wallet can house your payment cards — credit or debit — for use with Apple Pay and how to change between cards if you like. In this section, I cover other things you can do with the Wallet app.

After all, accessing Wallet from your wrist is awesome. For example, you can use it to have an airline attendant scan a QR code on your wrist to board a plane while you're negotiating an important deal on your iPhone. Plus, location and time services allow Wallet to notify you about relevant information just when you need it, such as gate changes at the airport.

When you download an app that supports Wallet, you have the option to add it to Wallet. Then, in your Wallet, you can flip through these apps, represented by virtual cards, with your fingertip to see each of their services and get them ready for when you need them. You can also be notified when you step inside of a supported retailer or another location, or when information regarding a particular service updates, such as a cancelled concert.

How to get going?

Wallet is already preinstalled on all iPhones. If the app you download supports it, Wallet asks you if it can add that particular app to your Wallet. Examples of supported apps include United Airlines, Target, Ticketmaster, Fandango, Walgreens, and Starbucks.

For example, to use the Starbucks app, follow these steps on your iPhone:

1. **Download the Starbucks app from the App Store.**

2. **Sign up for the Starbucks Rewards loyalty program or sign in if you're already a member.**

   You're prompted with this message: "Would you like to add your Starbucks card to your Wallet?"

3. **Tap Continue to add Starbucks to Wallet (Figure 10-7).**

   This message appears: "Choose your favorite stores, and your Starbucks card will appear on your lock screen when you arrive."

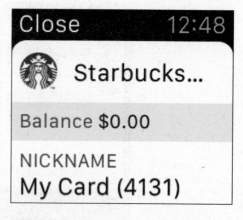

FIGURE 10-7:
Arguably, Starbucks is one of the more popular Wallet-supported loyalty cards. This is what the Starbucks app looks like on your Apple Watch.

4. **Let the phone's GPS locate the nearest Starbucks, and then you can select a few as favorites.**

   When you're done, you should see the Starbucks card pop up in your Wallet app — with a scannable QR code and cash balance of your account. Now hold out your Apple Watch to have the screen scanned by a barista at the cash register.

Using Apple Pay Cash on Apple Watch

So far, this chapter has discussed Apple Pay and how to use your Apple Watch or iPhone to make payments at retail locations or online. Did you know you can also send people money through your Apple device?

Called Apple Pay Cash, this easy-to-use technology is designed for person-to-person payments. It's essentially a version of Apple Pay that lets you send or receive money right in the Messages app (discussed in Chapter 5), or by asking Siri.

You can use the cards you already have in Wallet to send money, split a bill at a restaurant, or chip in on a gift for a coworker. Figure 10-8 shows this wonder in action.

FIGURE 10-8:
Available in the U.S. only, Apple Pay Cash lets you send or receive money via the Messages app.

Apple Pay Cash requirements

Here's what you need to get going:

>> United States residency. You also need to be 18 years of age.

>> A compatible device with iOS 11.2 and later or watchOS 4.2 and later.

>> Two-factor authentication for your Apple ID, which requires both a password and one-time code sent to your iPhone to confirm it's really you.

>> Connection (sign in) to iCloud with your Apple ID on any device that you want to use to send or receive money.

>> An eligible credit or debit card in Wallet (see earlier in this chapter on adding cards).

>> Agreement with the Terms and Conditions, something required the first time that you try to send or receive money.

>> Verification of your identity (you may be asked this).

Now, right from your iPhone, iPad, or Apple Watch, friends can securely send and receive money. Just be aware that Apple Cash is for personal use only and not for any business-related activities, such as paying business expenses or paying employees, per Apple.

Setting up Apple Pay Cash in Wallet

If you're setting up Apple Pay Cash for the very first time, you'll need to use any supported device that allows you to sign in to iCloud with your Apple ID. To do so:

1. Tap Settings ⇨ Wallet & Apple Pay.

2. Tap the Apple Pay Cash card, then follow the onscreen instructions.

3. To enable Apple Pay Cash on Apple Watch, open the Apple Watch app on iPhone, then tap My Watch ⇨ Wallet & Apple Pay (see Figure 10-9).

4. Now tap the Apple Pay Cash toggle so it turns green and says "Enable sending and receiving money in Message on Apple Watch."

FIGURE 10-9:
A look at
setting up
Apple Pay Cash
on iPhone (left)
and enabling
Apple Watch
support (right).

TIP

If you turn off Apple Pay Cash for any one device, such as on an iPad, you can still use Apple Pay Cash on other devices where you're signed in with your Apple ID, such as on an iPhone or Apple Watch.

Apple confirms there is no fee to use Apple Pay Cash with a debit card. If you send money using a credit card, though, there's a standard three percent credit card fee on the amount.

If someone sends you money via Apple Cash, it's automatically kept on your Apple Pay Cash card. You'll see your new Apple Pay Cash card in your Wallet, and you can use the money to send to someone; to make purchases using Apple Pay in stores, within apps, and on the web; or to transfer it from Apple Pay Cash to your bank account.

Using Apple Watch to send cash to friends

Once you're all set up, it's super easy to send cash to someone through your Apple Watch (see Figure 10-10)!

1. **Open the Messages app (see Chapter 5 for more on messaging).**

This may require you to press Digital Crown to see all your apps.

2. **Start a new conversation or tap to open an existing one.**

3. **Scroll down, then tap Apple Pay.**

FIGURE 10-10: Sending money to someone through Apple Watch is super easy, fast, and secure.

4. **Choose an amount by turning the Digital Crown.**

   To enter an exact amount, tap the dollar amount, tap after the decimal, then turn the Digital Crown.

5. **Tap Pay.**

   To review the payment information or cancel, scroll down. Your money in Apple Pay Cash will be used to pay first.

6. **To send the payment, double-press the side button.**

7. **To cancel a payment after sending, tap the payment to see its details, then check the Status field and tap Cancel Payment.**

TIP

You can use Siri to send money! Say something like, "Apple Pay 50 dollars to Steve for movie tickets" or "Send 25 dollars to Susan."

Using Apple Watch for Other Deals and Rewards

Besides using Apple Watch to make purchases and offer support for loyalty cards, you can also use it in ways that serve up a richer shopping experience. Folding in other technologies, you can utilize Apple Watch to customize the store experience to suit individual preferences. For example, stores equipped with Apple's iBeacon location-based technology (`developer.apple.com/ibeacon`) can identify you from your Apple Watch (and/or nearby iPhone). This means the technology welcomes you, rewards you with points for stepping into a store, notifies you about deals, and gives personalized suggestions based on previous interests.

Of course, these are things you have to opt into. That is, you must first give the store permission in its downloadable app.

Imagine stepping into a retailer and seeing what you need to know on the Apple Watch screen. It's still early days, but the technology is there. Just ask Cyriac Roeding, the celebrated entrepreneur, investor, and cofounder/CEO of Shopkick, an app that helps you find deals and earn rewards.

"While it's still a little early, smartwatches like Apple Watch are a next, natural step in the evolution of the shopping experience," (believes Roeding, whose Shopkick app is the most widely used shopping app at retail stores in the United States, according to Nielsen). As of April 2017, Shopkick said it had powered more than "200 million store visits, over 270 million product scans in aisle, and over \$2.5 billion in total sales from brand and retail partners" (according to Wikipedia). Roeding says Shopkick's Apple Watch app offers much of the same experience as the smartphone app "but without you having to take out your phone." When you walk into one of the many stores with Shopkick's shopBeacon transmitters — built upon Apple's location-based iBeacon technology — you're welcomed by name, rewarded with redeemable *kicks* (points) for walking in the door, and notified of curated deals based on your previous shopping habits. See Figure 10-11 for an example of Shopkick's technology (courtesy of Shopkick).

FIGURE 10-11: A look at Shopkick's shopBeacon transmitter on the wall of a retailer.

shopBeacon's location technology, which uses ultrasound and Bluetooth low energy (BLE), does not collect any information about you, assures Roeding, because it only sends information instead of collecting data.

4 More Apple Watch Tips and Tricks

Customize your Apple Watch by adding third-party apps, such as ESPN, Twitter, TripAdvisor, NPR One, and Fandango. You can now download apps directly to Apple Watch over the air.

Learn how to sync photos and download apps to your Apple Watch as well as how to establish and modify settings in the companion Apple Watch app on your iPhone.

Use Apple Watch to unlock your Mac, or to view verification codes on Apple Watch.

Discover how to turn your Apple Watch into a camera, including storing and viewing photos from your iPhone, as well as using the watch as a remote control for the iPhone's camera.

Enhance your Apple Watch experience by downloading gaming apps, including such iPhone favorites as Trivia Crack and Best Fiends.

» **Downloading and installing apps onto Apple Watch**

» **Discovering settings in the iPhone's Apple Watch app**

» **Using Apple Watch as an extra security tool for Mac owners**

» **Playing around with recommended third-party apps**

Chapter **11**

App It Up: Customizing Apple Watch with Awesome Apps and More

No two people are exactly alike, so why should our devices have the same apps installed?

Just like it's fun to customize a smartphone or tablet screen with all kinds of hand-picked apps — short for "applications," which is just a more trendy way to say "software" or "programs" — Apple Watch supports many thousands of third-party (non-Apple) apps.

The App Store for Apple Watch is already jammed with apps designed for your sleek wearable, including many familiar apps (perhaps based on its iPhone or iPad brethren) and completely new apps looking to dominate this new screen in your life. In most cases, they're free to download (or close to it), and you're sure to find something that resonates with you.

Bottom line: We're all individuals who have different tastes, priorities, and interests.

REMEMBER

You can now download apps directly onto Apple Watch, instead of requiring an iPhone.

This chapter looks at how to choose and install Apple Watch apps, and manage them from your Home screen. I take you through the companion Apple Watch app for iPhone and how to tweak settings and personalization options. Toward the end of this chapter, I also suggest several stellar apps you might want to consider for your watch. (See Chapter 12 for a brief discussion about gaming apps for your Apple Watch.)

Downloading Apps for Apple Watch

Because of the popularity of Apple's App Store, perhaps it's no surprise developers want to capitalize on watchOS and the Apple Watch. After all, millions of customers are wearing this gadget on their wrists and will likely want to customize their experience with apps.

In case you weren't aware, developers make 70 percent on paid apps, with the remaining 30 percent going to Apple. Thus, an opportunity exists to make some serious cash by creating a must-have app for Apple Watch.

"The more you wear Apple Watch, the more you'll realize just how personal a device it is," says Apple on its website. "Because with so many different apps available, you can choose the ones that are most relevant to you and create a customized experience.

"There are already apps for airlines, department stores, social networks and more that take advantage of the unique opportunities the wrist brings. And with new apps being built for Apple Watch every day, this is just the beginning," adds Apple.

How does one go about downloading apps?

From the App Store, but not the same App Store as the one on your iPhone, iPad, iPod touch, or Mac. It's a separate store solely dedicated to Apple Watch, as shown in Figure 11-1.

As previously mentioned, there are two ways to download apps to Apple Watch.

>> Use the Apple Watch App Store built into the Apple Watch app on your iPhone.

>> Download apps directly to Apple Watch (requires watchOS 6 or newer).

FIGURE 11-1:
A look at the
App Store for
Apple Watch,
which is part
of the Apple
Watch app on
iPhone.

From your iPhone

To use the Apple Watch App Store on your iPhone, follow these steps:

1. **On your iPhone, tap the Apple Watch app.**

   The black icon with a silver Apple Watch is preinstalled on all iPhone. Tapping the app launches the App Store for Apple Watch.

2. **Tap the App Store tab along the bottom of the screen.**

   You'll see some recommended apps to get you going.

3. **Find apps to download based on what's featured or by category.**

   For example, you can tap the See All tab beside Get Started to open a list of popular apps. Alternatively, tap the Search tab (as shown in Figure 11-2) in the lower right of the iPhone screen if you'd prefer to look for something by keyword instead of browsing.

FIGURE 11-2:
Use keywords, such as "instagram," to find new content for your Apple Watch.

Directly to your Apple Watch

To use the Apple Watch App Store on your Apple Watch, follow these steps:

1. **Tap the App Store iconon Apple Watch to launch the store.**

2. **If you haven't done so already, sign into the App Store. Tap Password and use the virtual keyboard (on a nearby iPhone) or Scribble feature to type it (only required once).**

   You might also be prompted to add a passcode to your Apple Watch to download apps, if you haven't already set that up.

3. **Browse through all the content to find something to download and try.**

   Search using Scribble or Dictation, or rely on your Siri voice assistant. Read reviews, look at the overall star rating, and check out screen grabs.

After it downloads, the app is available on your Apple Watch Home screen. You may be prompted first with a message like "Double click [side button] to install" or "Requires iPhone to complete setup." Just follow along! See Figure 11-3.

Many of the Apple Watch apps are free, and some require a nearby iPhone to get the most from them. Just remember that most Apple Watch apps have a companion app for iPhone; therefore, you need to have enough storage on your iPhone — not just on the watch — for many of them.

While at the App Store on your Apple Watch or iPhone, it's a good idea to read the full description of the app and check out the screenshots before you download it to ensure it's for you before you waste your time, data, and, perhaps, money. Also, read reviews by those who've already used the app.

TIP

You can also update the Apple Watch's operating system directly on the watch. Tap the Settings app icon, followed by Software Update, and if there's a newer operating system to download, it prompts you to download the update.

As most Apple Watch apps also install an iPhone version, after you download it, it syncs between devices. In other words, you don't need to do anything to copy apps over to the watch (if you downloaded Apple Watch apps to iPhone) or vice-versa (if you downloaded to Apple Watch directly). The moment the Apple Watch and iPhone are synced over Wi-Fi or Bluetooth, whatever you downloaded is transferred to your wrist.

FIGURE 11-3:
You can
download
Apple Watch
apps directly
to your wrist.
Here's what
they looks like
at the mini
App Store.

Deleting apps

To manage the apps on your Apple Watch — to delete ones you no longer want, for example, or perhaps to tweak the layout of them — you can do this in one of two ways.

>> **Use the Apple Watch app on your iPhone.** Simply launch the Apple Watch app on your iPhone and then tap My Watch, where you can uninstall apps you no longer want on your Apple Watch.

>> **Delete apps directly off Apple Watch.** From your Apple Watch's main Home screen — where you see all the circular icons in "Grid" view — simply press and hold lightly on an icon and they'll all start jiggling. If the app has a little "X" in the corner, you can tap it to delete it off Apple Watch. See Figure 11-4.

FIGURE 11-4:
You can delete third-party apps from Apple Watch itself. Tap the little "X" to remove.

If the app doesn't have a little "X," which is most (but not all) of the built-in Apple Watch apps, you can't delete them.

TIP

If you press too firmly on an Apple Watch app, you will see a prompt that lets you choose between the default "Grid" layout of the apps, or a "List" view. So, make sure you press lightly!

TIP

Unless you have a generous data plan, only download large apps on your iPhone when you're in a Wi-Fi hotspot. A couple megabytes is fine if you're using cellular data, but you should use wireless broadband for anything larger than that. Some games can be a couple gigabytes in size (1 gigabyte is roughly 1,000 megabytes).

Adjusting Settings in the Companion Apple Watch App on Your iPhone

You have tons of settings to change on your Apple Watch app. From changing the orientation of your watch to match your dominant hand to changing how your watch looks and sounds, the following sections cover all the details that help you make your watch yours. In this section, you also see personalization, general, accessibility, and first–party app options.

Settings and personalization

Instead of tapping the App Store tab in the lower right of the Apple Watch app on iPhone where you download Apple Watch apps, tap the My Watch tab in the lower left to open the settings and personalization options for the watch.

From within My Watch, you can perform the following actions:

>> Change the orange **App Layout** of the apps on your Apple Watch's Home screen.

>> Near the top of the screen, where you can see what **Apple Watch** you have (such as 44mm Case - Aluminum), you can **unpair** your Apple Watch from your iPhone.

>> You'll see the various **Watch/Clock Faces** illustrated at the top of the screen. Tap to select one.

>> Select **Complications** to change what info you see on your favorite clock faces.

>> The **Notifications** tab lets you customize what notifications you want pushed to the watch.

>> Select **Dock** to see (and customize) what apps you want to see when you press the side button.

>> Adjust brightness and text size for the watch in the blue-and-white **Brightness & Text Size** section. As shown in Figure 11-5, use the slider to adjust each setting and choose to bold text or not.

>> Enable or disable sounds and vibrations in the red-and-white **Sounds & Haptics** area (also shown in Figure 11-5). You should have options for Alert Volume, Mute, Cover to Mute (covering your watch with your hand to mute it), Haptic Strength, and Prominent Haptic for more pronounced vibrations.

>> The purple **Do Not Disturb** tab turns off all alerts, sounds, and haptics to the watch, whereas the orange **Airplane Mode** disables the wireless connection between Apple Watch and the iPhone. When enabled, the **Mirror iPhone** option automatically turns off Apple Watch's wireless radios whenever you do the same on the iPhone.

>> You should also find options for **Passcode** (enable a passcode on Apple Watch and select the four-digit code); **Health** (your birth date, sex, height, and weight for the purposes of estimating your calories burned); and **General**.

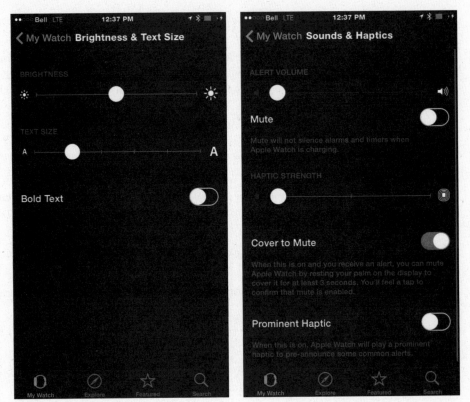

FIGURE 11-5: Use your fingertip to adjust settings for brightness, text size, sounds, and haptics.

General options

Under General options for Apple Watch, you should find the following information:

>> **About:** Software version installed and other information.

>> **Software Update:** To look for and download the latest watchOS update.

>> **Automatic Downloads:** For watchOS and app updates.

>> **Watch Orientation:** Left or right wrist and whether the Digital Crown button is on the left or right side. You have the ability to change this, as shown in Figure 11-6.

>> **Accessibility:** Options/aids for seeing and hearing impaired and more.

>> **Language & Region:** Choose the System Languages, Region, and Calendar type (which by default is Gregorian).

>> **Apple ID:** The account you're signed in to.

>> **Enable Handoff:** When this is on, your iPhone picks up where you left off with supported apps on your Apple Watch.

>> **Wrist Detection:** When this is on, Apple Watch automatically shows you the time and latest alerts when you raise your wrist.

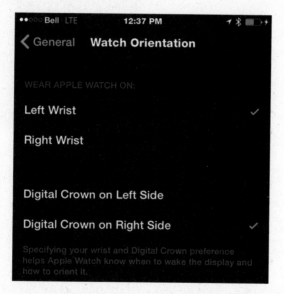

FIGURE 11-6:
You can change the watch's orientation from left wrist (default) to right and which side the Digital Crown button is on.

Accessibility options

Those living with visual or aural impairments can take advantage of a number of accessibility options on Apple Watch. Just as Apple has done with its iOS and OS X products (and the Safari web browser), Apple's smartwatch provides several features you can enable in the Accessibility area of the Apple Watch app on your iPhone, as shown in Figure 11-7. More details are available at Apple's website for these options (www.apple.com/accessibility/watch), but the following provides a quick summary:

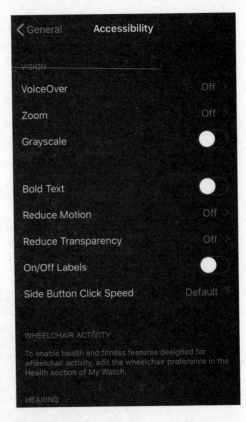

>> Visually impaired users can enable **VoiceOver**, a screen reader available in one of several languages.

>> Enlarge the Apple Watch fonts in **Font Adjustment** (ideal for the Mail, Messages, and Settings apps), add **Bold Text**, and opt for the **Extra Large Watch Face** option with bigger numbers to read.

>> Other visual tweaks include **Zoom** (magnification), **Grayscale** (no color), **Reduce Transparency** (increases contrast), **On/Off Labels** (easier to see if setting is on or off), and **Reduce Motion** (the Home screen is easier to navigate).

>> For those with impaired hearing, Apple Watch lets you enable **Mono Audio** (sending both audio channels to both ears and letting you adjust balance for greater volume in one ear or the other).

FIGURE 11-7:
You can adjust a number of accessibility options for Apple Watch, as shown in the Apple Watch app for iPhone.

>> Apple Watch's Taptic Engine can be enhanced with **Prominent Haptic**, a setting that gives a slightly more noticeable vibration and "preannouncement" for some alerts, such as calendar appointments, messages, phone calls, and more. A *preannouncement* means the watch sends you an additional haptic vibration before the main one.

You can also adjust some accessibility options on Apple Watch itself by triple-tapping the Digital Crown button.

REMEMBER

Security options

One of the watchOS 6 updates is the ability to leverage your Apple Watch as an extra security tool for Mac users.

As shown in Figure 11–8, the Mac includes an option to approve security prompts by tapping on the side button of Apple Watch, when prompted, to confirm it's really you logging in! Enable this option via the Mac's System Preferences ⇨ Security & Privacy area.

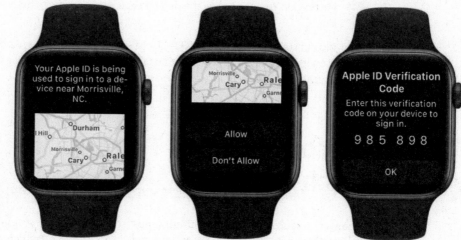

FIGURE 11-8:
Your Apple Watch can help log you into a Mac or display a one-time-use code to log in somewhere.

You can also use Apple Watch to display Apple ID verification codes, for when you need to log in to your Apple account on a new device or browser ("two factor authentication" combines something you know, such as a password, with something you have, such as Apple Watch).

First-party app options

Of course, multiple options and settings can be explored in each of the first-party apps installed on Apple Watch, such as: Activity, Calendar, Clock, Mail, Maps, Messages, Music, Passbook & Apple Pay, Phone, Photos, Reminders, Stocks, Weather, and Workout. Some examples of what you can change include:

>> **Clock:** You can select 12- or 24-hour (military) time; Push Alerts from iPhone (if enabled, Apple Watch lets you know about iPhone alerts); Notifications Indication (if enabled, a red dot appears at the top of the watch face when you

have unread notifications); Monogram (choose a one- to four-letter mono-gram to appear as a monogram complication on the Color watch face); and City Abbreviations (specifying different acronyms for the default ones provided for your chosen cities for the World Clock complications).

>> **Activity:** You can choose to turn on or off Stand Reminders (notified 50 minutes into an hour of inactivity); Progress Updates (set a time interval to receive an update on your Activity progress, such as every five hours); Goal Completions (on or off); Achievements (on or off); and Weekly Summary (on or off — received every Monday).

>> **Messages:** You can see all the default replies or change them to anything you like.

>> **Music:** You can select Synced Playlist (this playlist syncs when Apple Watch is on its charger) and Playlist Limit (such as changing it from 2 gigabytes to 1 gigabyte).

Twenty Recommended Third-Party Apple Watch Apps

Chapter 3 looks at all the default apps preinstalled on Apple Watch — you know, the ones Apple put there and you can't remove (which understandably frustrates a number of people) — but I want to share a number of optional but recommended and free third-party apps you can download from the Apple Watch App Store.

All of them take advantage of Apple Watch's features. Because I reserve games for Chapter 12, these 20 apps are tied more to information, travel, automotive, productivity, social media, fitness, shopping, and some entertainment.

Mint

The Apple Watch app for this popular finance tool lets you view your monthly spending goals at a glance as well as track your progress toward meeting them. And for those trying to stick to a budget, you can choose to receive weekly alerts with insight on how well you're doing (or not).

ESPN

One of the most popular sports apps for iPhone is now available for Apple Watch. Select which sports matter to you — such as baseball, football, basketball, hockey, golf, or tennis (or all the above) — and stay up to date with breaking sports news, real-time scores, and more, as shown in Figure 11-9.

FIGURE 11-9: The ESPN app should be a good fit for Apple Watch–wearing sports fans.

Target

As one of the first retailers to support Apple Watch, Target has an app that lets you build and view a shopping list on your watch so you can glance down to see what items you need — even if your phone is tucked away in your purse or pocket. When you enter a store, the Target app also tells you where to find the items you're looking for.

OneDrive

Although once bitter rivals, Microsoft has embraced Apple's iOS platform — and now the watchOS too. Based on Microsoft's OneDrive cloud service, this Apple Watch app lets users see their stored photos on their wrist — even when an iPhone isn't nearby.

SPG

As shown in Figure 11-10, one of the cooler apps is from Starwood Hotels & Resorts, which lets you unlock your hotel door by waving your Apple Watch at the sensor. A room key isn't required. The official SPG (Starwood Preferred Guest) app can also provide directions to your hotel, check you in, show your Starpoints balance, and more.

FIGURE 11-10: The SPG app lets you open hotel doors in select Starwood hotels, and much more.

Chirp for Twitter

While Twitter mysteriously pulled its Apple Watch support in 2017, you can see your Twitter feeds right on your wrist — with the help of a third-party app, Chirp for Twitter. And because they're only a couple of hundred characters, tweets fit perfectly on Apple Watch's small screen. Feel a gentle tap whenever new tweets are posted; plus you can retweet and favorite tweets from your Apple Watch. You can also send and receive direct messages, see the latest rrending topics, or search by hashtag. See Figure 11-11.

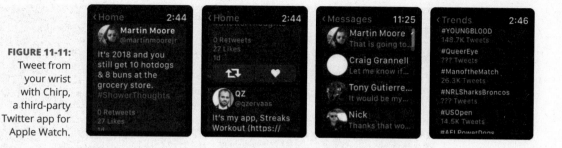

FIGURE 11-11:
Tweet from your wrist with Chirp, a third-party Twitter app for Apple Watch.

OpenTable

Hungry? The OpenTable app supports Apple Watch, which lets you see information about your upcoming dinner reservations by simply looking down at your wrist. The app can also help guide you to the restaurant with turn-by-turn directions.

Evernote

A popular productivity tool, Evernote for Apple Watch lets you view your stored notes, dictate a new one, set reminders, and search by keyword if you're looking for something in particular. Because Evernote stores your notes in the cloud, you can view your dictated notes in other Evernote apps — perhaps on a smartphone, tablet, or laptop.

American Airlines

How do you know when it's time to leave for the airport? Or if your flight has been delayed, cancelled, or changed gates? American Airlines (AA) has an Apple Watch app that can alert you to any and all of these things. The AA app also lets you check in for your flight, view a map with your estimated time of arrival, view baggage claim and connection details, and more, as shown in Figure 11-12.

BMW i Remote

Own an electric BMW i vehicle? The official Apple Watch app lets you remotely check on the charge status or notifies you when your car has been fully charged and is ready to go. This smartwatch app also lets you check your miles (to prevent "range anxiety"), see door-lock status, get service reminders, and view your cabin temperature.

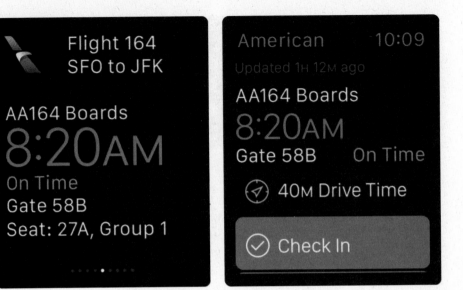

CNN

News junkies, rejoice! The official CNN app for Apple Watch gives you the information you need wherever life takes you. Select to receive breaking news and developing stories based on 12 categories of interest — such as Top Stories, U.S., World, Politics, Health, Entertainment, Sports, and Technology — plus your watch can even launch CNN TV live on your iPhone.

eBay

The world's largest marketplace is now a tap away. eBay on Apple Watch helps you keep up with the auctions you're watching — whether you're bidding on something or selling merchandise. The app conveniently lets you send and receive alerts without having to fumble through your phone, tablet, or personal computer.

Citymapper

If you rely on public transit, the Citymapper app for Apple Watch always shows you the best bus and train routes based on your location and where you want to go. You should see step-by-step instructions, including a list of the next three arrival times for your mode of transportation so you can decide when to leave, and you should feel a vibration on your wrist when it's time to get off at your stop. See Figure 11-13 for a look at the Citymapper app.

FIGURE 11-13: Ideal for those who take public transit, the Citymapper app for Apple Watch shows you relevant information.

TripAdvisor

Find that hidden gem of a restaurant on your next trip. Unearth dozens of things to do while discovering a new town. Everything that makes TripAdvisor the perfect travel companion is now on your Apple Watch. Get instant information on hundreds of nearby restaurants, sights, and tourist destinations.

NPR One

Fans of NPR can make their favorite station even more personal. The NPR One app shows you relevant news and curated stories based on your interests, along with access to your playlist (on your iPhone), and you can search for specific shows by using dictation and control basic playback functions with your fingertip.

Fandango

The popular movie ticketing app is now on your wrist. After you've purchased tickets to a flick, the Apple Watch app can display the movie time and theater location, phone number, and other information you might need.

Things

If you wear a watch — and a smartwatch, no less — it might not be a stretch to assume you like to be organized. But that doesn't mean you're good at it! If you need a little help, an app called Things is an excellent to-do manager for iPhone, and it supports Apple Watch too. Organize your life with daily tasks, which you can easily sort into sections like Today, Upcoming, and Anytime. Specifically, the Apple Watch app focuses on your current tasks, which can be displayed as a watch face complication or in the app itself, and lets you tick off items when completed.

PayByPhone Parking

You can use the PayByPhone Parking app on Apple Watch to pay the meter, check on the time remaining, and deliver an alert ten minutes before the meter expires. If you're not done with your errands, your watch lets you add more time to the meter without your having to go back to your car.

Sky Guide

Watch the skies! And your wrist. As shown in Figure 11-14, the Sky Guide app for Apple Watch is great for armchair astronomers. Receive alerts about upcoming celestial events — such as meteor shows and eclipses — and it even alerts you when the International Space Station is about to fly over your location.

Lutron Caséta

Your smartwatch can control your smart home. The Lutron Caséta app for Apple Watch lets you control the lights in your home even when you're not there so you can make it look like you're home when you're on vacation. Or on the flipside, if you accidentally leave the lights on when you leave, you can get an alert on your wrist to turn them off.

FIGURE 11-14:
The Sky Guide app is out of this world (badum bum!). Read about astronomical events and receive alerts about them too.

IN THIS CHAPTER

» **Looking at photos on Apple Watch**

» **Syncing photos to Apple Watch**

» **Using your watch as a shutter button for iPhone**

» **Playing around with 10 recommended games for Apple Watch**

Chapter **12**

Extra! Extra! Having Fun with Apple Watch

You didn't think Apple Watch was just for information, communication, and navigation, did you?

This mobile companion of yours is also ideal for playing around. In Chapter 5, I look at how to send animated emojis and digital sketches to people. And Chapter 9 covers music and podcast playback, audiobooks, and how to control Apple TV playlists from your wrist.

But your watch can do much more in the fun department.

Granted, Apple doesn't seem to advertise these nonessential applications as much as customizing watch faces, sending messages, or calculating your physical activity, but you can indeed enjoy some downtime with Apple Watch, including many playable games already available in the App Store for Apple Watch.

You can also look at photos of people, pets, and places on your Apple Watch — anytime and anywhere. Perhaps you ran into someone who asked how old your daughter is. Now you can show that person her smiling face. Or maybe you want to glance down to see old friends when a song on the radio brings back camp memories. If you're feeling like you need a vacation, call up photos of last year's trip to Jamaica so you can see the white sand and blue water (and then call your travel agent!).

Speaking of photos, Apple Watch can let you access your iPhone's camera to snap the shutter button wirelessly, which is ideal for selfies and group shots. This is available through the Camera app.

Copying Photos to Apple Watch

If you own an Apple Watch without cellular access, you likely know most tasks require a nearby iPhone. The two devices are wirelessly tethered via Bluetooth and Wi-Fi, but you can sync some files to Apple Watch just in case your iPhone isn't nearby.

Chapter 9 talks about syncing music to the watch — perhaps if you want to go for a jog around the neighborhood without your phone and you own a Bluetooth headset — but here, I cover transferring photos over to your watch.

I'm not exactly sure *why* you'd want to copy photos to your watch, unless you really think you're going to want to see some pictures when your iPhone isn't nearby. Having music on the watch makes more sense — per my jogging scenario — but those who want to take advantage of this feature must first enable it on the Apple Watch app on iPhone.

To copy photos onto your Apple Watch, follow these steps:

1. **Open the Apple Watch app on your iPhone and tap My Watch in the bottom left of the screen.**

2. **Swipe down until you see the Photos icon and then tap it.**

3. **Some options appear, including one to mirror your iPhone (where photos are synced), or you can choose Custom to handpick what's synced to the watch.**

   As shown in Figures 12-1 and 12-2, you've got some options in the Photos area of the Apple Watch app on iPhone:

   - *Synced Album:* Select which iPhone-stored photos are viewable on Apple Watch, even when you don't have your iPhone with you. By default, it's your Favorites album, but you can also choose another one, such as Camera Roll. Or select None.

- *Photos Limit:* Select the photo storage limit on your Apple Watch. Although this limit is measured in megabytes (MB), you can raise or lower the number, but 15MB is the default number, which translates to 100 photos. Lowering it to 5MB loads only 25 photos. You can raise it to 40MB (250 photos) or a maximum of 75MB (500 photos).

Be aware you can also view photos from your iCloud account on Apple Watch.

REMEMBER

Interestingly, you can also copy up to 2 gigabytes of music to Apple Watch, which translates to roughly 500 songs. Therefore, you can store up to 500 photos and up to 500 tunes. You can't go over this maximum for photos and music.

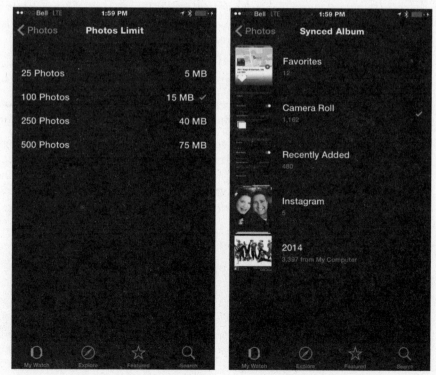

FIGURE 12-1:
Select which photos you want synced to Apple Watch (if any).

FIGURE 12-2:
Select how many photos you'd like to sync (by size or number of files) and from which album.

Launching Photos on Apple Watch

Okay, so Apple Watch doesn't have the biggest screen in your life, but it *is* always on your wrist; therefore, it's a conveniently placed digital photo frame.

One of the built-in apps is Photos, which is similar to a photo gallery app on your iPhone or iPad.

To use the Photos app on your Apple Watch, follow these steps:

1. **Tap the Digital Crown button to go to the Home screen.**

2. **Tap the Photos app.**

   If you prefer, raise your wrist and say "Hey Siri, Photos." Either action launches the Photos app, which shows you thumbnails of photos stored on your phone (or watch).

3. **Twist the Digital Crown button to zoom in and out on individual images.**

   Twisting the Digital Crown away from you zooms out to see more photos, which makes the thumbnails smaller, whereas twisting toward you zooms into a photo.

4. **Zoom in until a photo takes up the entire watch face.**

   Now you can swipe left and right to browse through your photos one at a time. The photos are in the same order as they are on your iPhone's Photos app, including any albums you've created. Figure 12-3 shows a photo full screen on Apple Watch.

   Why doesn't Apple Watch allow you to pinch and zoom? Your finger and thumb would cover up your photos. The Digital Crown button works better.

5. **Twist the Digital Crown button to zoom out to see more photos, or tap the Digital Crown button to exit the Photos app and return to the Home screen.**

   Speaking of photos, don't forget Apple Watch can show you images that are embedded/attached to Messages and Mail (email). You should see images just below the text in a given message. Twist the Digital Crown button to see the accompanying photo(s) near the bottom of the screen. Too bad you can't send a photo to someone from Apple Watch. For that, you need to use your iPhone.

DID YOU KNOW?

Speaking of photos, among the first celebrities seen wearing Apple Watch — even before it went on sale April 24, 2015 — were pop star Katy Perry and rapper Drake (both of whom were donning a yellow gold watch with a red leather band) as well as "Happy" singer/songwriter Pharrell (with white leather band). Being a celeb has its perks!

Discovering the Camera App

Did you know you can use Apple Watch as a viewfinder for your iPhone's iSight (rear-facing) camera? You can see a preview of your photo before you take it, set the camera timer on your watch, or just take the photo.

The Camera app might be ideal for those who want to get in the picture but don't want to press the iPhone's shutter button (on the screen or a button along the side of the phone). It can be awkward to hold the camera and take a selfie at the same time or, worse, risk dropping your iPhone and breaking the glass. Or you can use one of those selfie sticks too.

To use the Camera app on your Apple Watch, follow these steps:

1. Tap the Digital Crown button to go to the Home screen.

2. Tap the Camera app.

Alternatively, you can lift your wrist and say "Hey, Siri, Camera." Either action launches the Camera app. You should then see a preview of your iPhone's iSight camera.

3. **Frame up your shot by getting your subjects huddled together or centering a landscape photo or whatever, as shown in Figure 12-4.**

Consider Apple Watch your live viewfinder for the iPhone you're holding (or one you've placed on a tripod).

FIGURE 12-4:
No broccoli in the teeth? As you can see here, you can take a selfie or a group shot with friends.

4. **If you want to take a photo, press the white shutter button in the center of the watch — just below the preview window.**

This instantly snaps the picture and adds a thumbnail to the bottom left of your Apple Watch's screen. You can tap this thumbnail if you'd like to see the photo full screen.

5. **If you want to use the timer, tap where it says 3s (3 seconds) and then tap the white shutter button (shown in Figure 12-5).**

You should see a countdown on the screen before the photo is taken.

Review what you took, and if you like it, press the Digital Crown button to return to your Home screen. Don't bother pressing the white shutter button if you don't like what you see. Frame up a different shot and then press the white button.

FIGURE 12-5:
Whether you use the timer or not, tap the large white shutter button to snap a picture by using your iPhone's iSight camera. It counts down from 3 to 1.

A quick checklist of what you can do when viewing photos on the Camera app on Apple Watch:

- » **View a photo**: Tap the thumbnail in the bottom left.
- » **See other photos**: Swipe left or right.
- » **Zoom**: Turn the Digital Crown.
- » **Pan**: Drag on a zoomed photo.
- » **Fill the screen**: Double-tap the screen.
- » **Show or hide the Close button and the shot count**: Tap the screen. When finished, tap Close.

CAMERA-LESS SMARTWATCH — FOR NOW

Although Apple Watch can serve as a viewfinder for your iPhone's camera, letting you snap pictures wirelessly and even set the self-timer, it doesn't have its own built-in camera like some other smartwatches have. Shoppers should remember this if it's an important feature to you. Perhaps Apple will add it to future models?

Examining a Batch of Apple Watch Games

When smartphones started to take off in the early part of the 21st century — including iPhone's high-profile debut in 2007 — many video game purists said the screen was simply too small to provide a gratifying interactive entertainment experience. Why would people want to squint to play a game on a (then) 3.5-inch screen when they have a 14-inch laptop, 23-inch desktop, and 60-inch flat-panel TV at home? And don't you need a game controller with real buttons as opposed to finger taps, swipes, and flicks on a small screen?

But then came along a handful of stellar games that proved the naysayers wrong: Angry Birds, Temple Run, Flappy Bird, Subway Surfers, Cut the Rope, Candy Crush Saga, Fruit Ninja, Words With Friends, and Jetpack Joyride, to name just a few.

Gaming on a smartphone is not only convenient — because we don't go anywhere without these devices — but the App Store offers a ton of games, with most downloadable games costing just $0.99 cents, if anything at all.

Whether Apple Watch will be a viable gaming platform is still up for debate, and developers have a few obstacles to overcome — a tiny screen, limited user interface, and questionable battery life — but where there's a will, there's a way — and an app for that! Gamers love to play anywhere, anytime, and on any device, so "time" will tell whether the wrist will be the new place to game.

That hasn't stopped a number of game developers from creating digital diversions on Apple Watch, many of which were announced even before the watch became available in spring 2015.

The following is a look at 10 great games. Remember, to find and download these games and other apps, you can do it in two ways: open the App Store on your Apple Watch or launch the Apple Watch app on your iPhone and then tap the App Store tab in the lower-right corner of the screen.

Watch This Homerun

From Atlanta-based Eyes Wide Games comes the first of many sports games. Watch This Homerun is a watch-sized baseball experience, delivering bite-sized (10- to 15-second) game experiences. The game has you touch the screen at the right time to whack baseballs — be they fastballs, curveballs, or changeups — into the bleachers, with inaccurate taps sending you back to the bench.

The game's companion iPhone app tracks competitive leaderboards, achievements, and personal stats against friends and the rest of the community. Future one-touch sports games include football, basketball, soccer, and boxing, says the developer. See Figure 12-6 for a look at Watch This Homerun.

FIGURE 12-6: Knock a baseball into the stands with Watch This Homerun.

Trivia Crack

One of the biggest trivia games on iPhone is also available for Apple Watch. Trivia Crack (as shown in Figure 12-7) lets you play rounds of pop culture trivia — without ever taking your iPhone out of your pocket. Animated characters guide you in defeating all your opponents while you answer millions of questions submitted by users from around the world. Questions appear on the watch's screen, and you can answer them right on the watch; after an opponent answers and it's your turn, you're notified from your watch and can continue playing from there. Example: What's Muhammad Ali's real name? Rocky Balboa, Rocky Marciano, Anderson Silva, or Cassius Clay (correct answer).

Runeblade

Gamers looking for a richer gaming experience might consider Runeblade, a fantasy adventure RPG (role-playing game) that's simple enough for anyone to pick up and play but offers far more depth for players looking for a deeper experience, says Everywear Games. The company's first Apple Watch title increases in depth and complexity over days, weeks, and months of gameplay, although games are designed to be played in 5- to 15-second sessions. As shown in Figure 12-8, the game stars you as a High Priestess of the War Mages, challenged with stopping an ancient menace from consuming the realm as you use might and magic to defeat mystical creatures, unlock new spells and abilities, and level up to grow stronger over time.

FIGURE 12-7: This is what Trivia Crack looks like on Apple Watch's screen.

FIGURE 12-8: Screenshots from Runeblade by Everywear Games.

Best Fiends

If you like such matching games as Candy Crush Saga, Apple Watch owners might enjoy tapping through Best Fiends — one of the first mobile games for the platform and based on the popular iOS version. In the single-player puzzle game from developer Seriously, players explore the lands of Minutia, collect treasures, and battle the malevolent Slugs of Mount Boom (such as the one shown in Figure 12-9). What's more, players earn rewards that can be applied to their Best Fiends game on their iPhone — used toward leveling up characters, unlocking new powers, and defeating enemies. In the iOS version, players drag a finger up, down, and diagonally to match identical items on an obstacle-laden game board.

FIGURE 12-9: Best Fiends (not Best Friends) is a popular matching puzzle game that's now playable on Apple Watch.

Snappy Word

Australian developer Right Pedal has released Snappy Word, a word game that places four random letter tiles on the watch face. By tapping the letters in order to create words, your goal is to see how many words you can create in 30 seconds. For example, *E*, *M*, *S*, and *A* can create such words as *same*, *mesa*, and *as*. Also playable on iPhone, iPad, or iPod, this Apple Watch game (as shown in Figure 12-10) includes a real-time player-versus-player mode, or you can compete against others through the Game Center leaderboard. Snappy Word's dictionary is said to be made up of nearly 4,000 words.

Watch Quest

Even before Apple Watch debuted in the spring of 2015, buzz amassed in the gaming world over an adventure title called Watch Quest. Players first choose characters on the companion app for iPhone, equip them with gear, and then have them embark on a quest on Apple Watch — filled with engaging in combat, solving puzzles, and hunting treasure. Starring a male or female protagonist, this colorful game includes passive play — slower-paced and easier quests — and more difficult active play, including battling monsters (as shown in Figure 12-11) and foraging for items. Watch Quest is free to download and comes with one training campaign, but you have to pay for additional quests and characters from within the iPhone app.

Spy_Watch

One of the more intriguing offerings sneaking onto Apple Watch is from UK-based indie developer Bossa Studios — best known for Surgeon Simulator and I Am Bread.In Spy_Watch, as shown in Figure 12-12, you're the head of a spy agency that's seen better days. To turn the agency around, you must train spies, send them on secret missions, and earn money to improve their abilities.

FIGURE 12-11:
Fans of epic quests might enjoy this Apple Watch–supported adventure: Watch Quest.

According to the developer, this ambitious Apple Watch game centers around short bursts of real-time alerts between you and your agent in the field, which take place in two- to five-second rounds. As the agency head, you need to make strategic decisions on the fly that will ultimately affect the success of the missions — be they related to time, resources, or locations. No guarantees you'll look as suave as James Bond while tapping on your wrist.

Rules!

With the iOS version taking home multiple award nods and an official App Store pick, Rules! was one of the first games announced for Apple Watch. And it still remains popular today.

In case you haven't played the hit action puzzler from The Coding Monkeys, you're tasked with tapping tiles based on the rules for the level, such as "Tap numbers in descending order," "Tap green" (tiles), or "Tap animals" (opposed, to, say, monsters or robots). When you advance to a new level, the rules revert to an older one, and you must remember what they were. See Figure 12-13.

Along with its 100+ memory-testing levels, there's also a daily "brain workout" mini-game challenge, an animated sticker pack, multiple difficulty settings, and other noteworthy features.

FIGURE 12-12:
A stealthy spy
game for Apple
Watch called
(you guessed
it) Spy_Watch.

Wordie

Another great word game is just like other versions — on iPhone, iPad, and in iMessage — but is now playable on your wrist. ICO Group's Wordie is a fun game that shows you four photos and has you guessing the common word between them. See Figure 12-14.

For example, you might see a photo of people picking up trash in bags, a woman undergoing facial surgery, a ball pit at a playground, and a vinyl record. The answer here is PLASTIC, which you guess by the number of spaces available (like Hangman) and available letters to choose from.

FIGURE 12-13:
Rules! is an extraordinary puzzle game for Apple Watch.

FIGURE 12-14:
Ideal for Apple Watch gamers, Wordie! is an addictive puzzle game to play on the go.

There are more than 600 puzzles; plus you can create your own and share with friends.

If you're stuck on a level, you can also ask your friends for help on social media or use hints.

Cosmos Rings

From Square Enix, the same company that introduced Final Fantasy and Secret of Mana, comes a mobile adventure with charming pixelated graphics. This simple-to-pick-up yet surprisingly deep role-playing game (RPG) offers a unique combat mechanic designed for the small screen: while basic attacks are automatic, you must enable powerful skill relics, strengthen your sword, and initiate a chain sequence to battle your way through this dark world. See Figure 12-15.

Need to go back in time to retake a mission? Simply twist the Apple Watch's Digital Crown to turn back the clock!

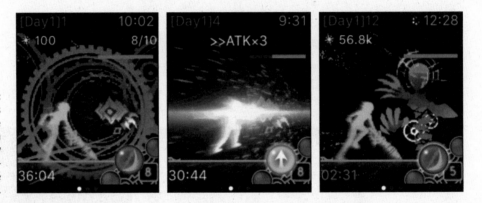

FIGURE 12-15: Cosmos Rings from Square Enix is a surprisingly deep role-playing game for your wrist.

5

The Part of Ten

IN THIS PART . . .

Discover the top ten things you should try with your Apple Watch as soon as you've taken it from its box and charged it up. These include turning your Apple Watch into a hotel key or a Walkie Talkie, sending personalized sketches to friends and family with Digital Touch, and, of course, learning more ways Siri can help you in your day-to-day life. You can also quickly learn how to play music on your Apple Watch, establish your fitness goals, and set up Apple Pay.

Chapter **13**

Ten Cool Things to Do with Your Apple Watch

I f you've spent some time flipping through this book — and thank you by the way — you've no doubt soaked up a lot of what Apple's "most personal device . . . ever" can do.

Apple Watch has hundreds of use-case scenarios — many of which are provided by Apple and its built-in features — while third-party apps extend the functionality of this smartwatch even further (as evidenced in Chapters 11 and 12).

But if you're like most people, you won't have time to go over *everything* Apple Watch has to offer. As I mention earlier in this book, it's estimated we only use about ten percent of what our gadgets can do — until someone shows you what you're missing.

As the author of *Siri For Dummies*, I always love that "a-ha!" moment when I teach people a Siri tip or trick they had no idea about. It's even more gratifying when they tell me at a later time that what I showed them is something they've grown to rely on in their day-to-day lives.

This brings us to this final chapter of *Apple Watch For Dummies*, where I isolate ten must-try features of Apple Watch. Consider it a purely subjective list of my favorite things about the device. Perhaps you don't know where to start or want to know how to best showcase your new gadget to friends and family. Or maybe you just want a refresher on a handful of the most useful features of Apple Watch — all in one chapter — instead of picking and choosing the individual tasks from across a couple hundred pages?

Thus, the following are my top ten features of Apple Watch and how to enable them. I also mention where in this book you can learn more about each one.

Activity

Many smartwatches and fitness bands can report on your performance while working out, but Apple Watch is always calculating what you're doing — or not doing.

The innovative Activity app and its three rings — for Move, Exercise, and Stand — does a stellar job of giving you an idea of your overall physical activity.

Press the Digital Crown button or lift your wrist and say "Hey, Siri, Activity" and then take a gander at your progress:

>> The reddish-pink **Move** ring shows how many calories you've burned by moving around during the day.

>> The lime-green **Exercise** ring is for minutes of brisk or intense activity you've completed that day.

>> The baby-blue **Stand** ring gives you a visual indication of how often you've stood up after sitting or reclining.

Your goal is to complete each ring each day. The more solid each ring is, the better you're doing. Plus, you can swipe around inside the app for a numerical look at your performance.

DID YOU KNOW?

You can multitask on Apple Watch. Although you're not seeing two apps on the screen at the same time, you can tap the side button to pull up your Dock to your last used apps (or if you prefer, your favorite apps). For example, you can toggle between the Music app and the Weather app to catch a multiday forecast, and then switch back to the Music app by flicking up or down and tapping the watch when you see what you'd like to see full screen. Because Apple thinks you might want to see the time more than other information, the clock is visible in many apps already; therefore, you don't need to leave the app you're in to see what time it is.

You can also change your goals per day — an example is shown in Figure 13-1 — in case they're too ambitious for your lifestyle, or you can bump them up for an added challenge. The companion Activity app for iPhone shows you additional information, including a historical look at your Activity levels. Plus, every Monday, you should receive a summary report on your Apple Watch about your activity and goals. And there's the new Trends feature, too, for an even deeper look at your activity history (synced with iPhone).

Chapter 8 offers a closer look at the Activity and Workout apps.

FIGURE 13-1:
You can change your daily goals depending on how ambitious you are.

Apple Pay

Using your watch to buy things at retail is incredibly convenient. Even if you don't have your iPhone around, you can wave your wrist over one of those contactless terminals at the checkout counter or at an Apple Pay–compatible vending machine and the transaction is completed. And securely.

To buy something using Apple Pay on your Apple Watch, follow these steps:

1. **Double-tap the side button on Apple Watch, which opens Apple Pay.**

    Apple Pay uses your default card in the Passbook app, but you can change it to something else if you like. See Chapter 10 for more on the Passbook app.

2. **Hold the watch up to the contactless terminal and you should hear a tone and feel a slight vibration — both of which confirm the payment has been made.**

   That's all there is to it. Apple Pay uses near field communication (NFC) technology inside of Apple Watch to make the *digital handshake* with the retailer's contactless terminal.

Apple Pay is supported by many banks and financial institutions as well as many thousands of retailers. But remember, you need to set up Apple Pay first on your iPhone if you haven't done so already. See Chapter 10 for more on Apple Pay.

Hotel Key

Free apps like Starwood Preferred Guest (SPG) let you tap your watch on your hotel door to gain entrance. No more fumbling for the key card or having it demagnetized because you had it in your pocket with your smartphone.

If you've got the free app installed, tell someone at the check-in desk at a Sheraton, Westin, W Hotel, Meridien, St. Regis, Element, or Aloft. Keep in mind that support for Apple Watch likely won't be available at all Starwood hotels and resorts or rolled out at the same time.

In the near future, expect many similar apps to let you into your car — instead of needing a large key fob — or to enter public transit stations, including bus depots and train terminals. Perhaps soon, Apple Watch will let you walk through your front door at home (with high-tech deadbolts, such as the Kevo from Kwikset) or into your office by tapping your wrist on a card reader. See Chapter 11 for other third-party Apple Watch apps to check out.

Walkie-Talkie

In case you haven't given it a shot yet, Walkie-Talkie is a fun way for two Apple Watch wearers to communicate between themselves. As you might expect, Walkie-Talkie lets you quickly chat with someone, wrist to wrist, using your voice (see Figure 13-2).

FIGURE 13-2:
Use the Walkie-Talkie feature to contact another Apple Watch wearer instantly. And it's fun!

To get going, you and the person you'd like to talk with need to set up the FaceTime app on your iPhone. This app enables you make and receive FaceTime audio calls. See Chapter 5 for into on that.

To use the Walkie-Talkie app on your Apple Watch, follow these steps:

1. **Open the Walkie-Talkie app (it's yellow and black) on your Apple Watch.**

2. **Press the yellow + sign and choose a contact.**

   Wait for your friend to accept the invitation. The contact card remains gray and reads "invited" until your friend accepts.

3. **After your friend accepts, his or her contact card turns yellow.**

   You and your friend can now talk instantly.

4. **Touch and hold the talk button, and then say something.**

   Now your friend can hear your voice and talk with you instantly.

5. **To talk over Walkie-Talkie, touch and hold the talk button, then say something; when you're done, let go.**

   Your friend instantly hears what you said. To change the volume, turn the Digital Crown.

REMEMBER

Apple Watch has Wi-Fi or cellular support on some models, but that doesn't mean you can surf the web; Apple Watch doesn't come with a web browser; it uses Wi-Fi only to move or sync data between it and your iPhone. That's probably not a bad thing, given the fact Wi-Fi eats up valuable battery life pretty quickly. Also remember that Walkie-Talkie requires that both people be running WatchOS 5 or greater.

Music Playback

Many people who exercise rely on music to help keep them entertained and motivated. You might not want to bring a large iPhone with you on a jog or run, so Apple thankfully lets you sync some music to Apple Watch — up to 2 gigabytes, or about 500 songs.

To sync music to your Apple Watch, follow these steps:

1. **Connect your Apple Watch to your PC or Mac via its USB charger.**

   Use the magnetic charger that shipped with your Apple Watch.

2. **On your iPhone, open the Apple Watch app.**

3. **Under My Watch, scroll down and tap Music, followed by Synced Playlist.**

   Decide what you'd like to transfer over to your watch: My Top Rated, Recently Added, Recently Played, Top 25 Most Played, or Purchased tracks.

4. **Tap to select one of these options.**

   Unplug the Apple Watch from the computer when the sync is complete.

After you have songs stored on your Apple Watch, follow these steps:

1. **Open the Music app and press and hold the screen (Force Touch) to launch a couple options.**

   You should see options for Shuffle, Repeat, AirPlay, and Device.

2. **Tap Device and then select Apple Watch rather than iPhone.**

   You should be prompted to pair a Bluetooth-enabled headset or headphones to hear the music. The Apple Watch screen shows you what's playing on your watch or iPhone.

Apple Watch also acts as a remote control for an Apple TV connected to a TV or for playing music on an iPhone or iPad. See Chapter 9 for more on these and other media- and entertainment-related tasks.

Maps

Because Apple Watch is always on your wrist, it's a conveniently placed screen for getting directions. Apple Watch can give you turn-by-turn directions by tapping into your nearby iPhone's GPS chip, and you should see the overhead map on your watch, including a blue dot for your location, a red pushpin for the destination, and the path to take to get there quickly. Apple Watch gently vibrates to tell you when it's time to turn left or right.

To use the Maps app on your Apple Watch, follow these steps:

1. **Press the Digital Crown button to go to the Home screen.**

2. **Tap the Maps app.**

    You can also raise your wrist and say "Hey, Siri, Maps." Either action opens the Maps app. An overhead map of your current location appears on the Apple Watch screen, and you can swipe in any given direction to move the map around, or you can twist the Digital Crown button if you want to see nearby streets or businesses.

3. **Press and hold the screen and then speak an address or business name.**

    If you make a mistake, tap Clear. If you're happy with what you requested, continue to the next step.

4. **Tap Start to begin the turn-by-turn directions.**

    You should now see and feel when it's time to turn left or right when nearing an intersection — whether you're on foot or in a vehicle. Your iPhone also shows you information if you want to peek at a bigger screen (safely) or hand it to a passenger. See Chapter 6 for more on the Maps app.

Digital Touch

Many smartwatches on the market offer similar features, such as seeing who's calling or texting, calculating fitness information, or getting directions to a destination. But Apple Watch offers a few unique watch-to-watch communication options — collectively referred to as Digital Touch (see Chapters 3 and 5 for more on this feature). Here are three examples of them:

>> **Sketch:** Draw something with your finger, and the person you're sending it to sees it animate on his or her Apple Watch.

>> **Tap:** Send gentle (and even customizable) taps to someone to let that person know you're thinking about him or her.

>> **Heartbeat:** Your built-in heart rate monitor is captured and sent to someone special so that person can feel it on his or her wrist.

To send a heartbeat with your Apple Watch, follow these steps:

1. **Open the Messages app and tap someone.**

   Start a new message or reply to an existing conversation.

2. **Tap the blue and white icon that looks like two fingers on a heart.**

   You see a black screen, ready for your fingers. And you can tap the top-right corner to change colors.

3. **Press and hold your fingers on the screen and you'll feel it pulse.**

   When you feel the pulsations stop, you can lift your fingers up and your heartbeat is sent to the recipient (shown in Figure 13-3).

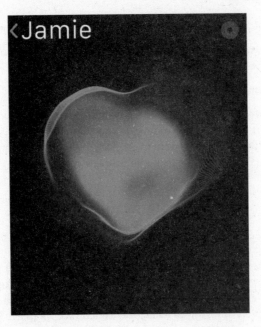

FIGURE 13-3: Want to tell a special someone you're thinking about them? A romantic way to do it is to send your heartbeat — but they'll need an Apple Watch to feel it!

Siri

Because Apple Watch was designed for quick interactions and to get information wherever and whenever you need it most, the best way to interact with your watch is by your voice. Providing you're in a place where you can talk freely, speaking into your watch's microphone is a fast, accurate, and convenient method for getting what you want when you want it.

If you recall, you can use Siri in two ways on Apple Watch:

>> **Digital Crown:** Press the Digital Crown button and wait to see the little bars jumping up and down near the bottom of your screen. This confirms Siri is "listening" to you.

>> **Voice activation:** Raise your wrist and say "Hey Siri," followed by your command or question. Or you can go into and change the Settings of Apple Watch to enable simply raising your wrist to activate your personal assistant.

For both of these options, you should get what you need within a second or two, but remember, you need your iPhone nearby because your request is quickly sent to Apple's servers to process it.

Siri can help you with virtually any task, including some of the following, which are tied to various apps and online content:

>> "What time is it in Milan?"

>> "Read me my messages."

>> "Text Susan that I'll be five minutes late."

>> "Call Dad."

>> "Show me my email."

>> "When is my next appointment?"

>> "Open the Activity app" or "Open the Workout app."

>> "Where is the closest gas station?"

>> "What song is playing?"

>> "Play jazz."

>> "What's the weather going to be like tomorrow?"

>> "How are the Braves doing?"

>> "Set an alarm for 6 a.m."

See Chapter 7 for more on using Siri to help you complete tasks with your Apple Watch.

Watches and Watch Faces

Most companies that release a smartwatch have one or two models, but Apple Watch is available in multiple sizes, several materials (aluminum, stainless steel, and 18-karat gold), in multiple case colors, and with various band colors, materials, and styles to choose from.

Clearly, Apple has thought this through!

Even with all the options, the user experience will be similar between all the versions because features, interfaces, and apps are the same for all of them.

The following is a quick summary of the five current options (at the time of writing this second edition of the book):

>> **Apple Watch Series 3:** Although not the newest Apple Watch available, this model was the first to offer either a GPS chip — to accurately capture location information — or a GPS and cellular option. More on this later.

>> **Apple Watch Series 4:** Available in late 2018, these Apple Watch options have slightly larger screen — but in the same-sized body — and offer additional features including an ECG (electrocardiogram), fall detection, and more.

>> **Apple Watch Series 5:** This Apple Watch stays on all the time, thanks to a new type of screen (but dims when you don't need it). It also adds an integrated compass, new materials (like ceramic), personalized complications (such as activity levels, weather, and UV index), and support for international calls to emergency services (with the cellular version).

>> **Apple Watch Nike+:** Ideal for fitness types who like the Nike brand, this special edition Apple Watch Series 4 watch (and special loop band) was designed to be your running partner and synchronizes with the Nike Run Club app and Nike Training Club app.

DID YOU KNOW?

Some people — nay, many people — thought Apple was crazy to create an 18-karat-gold version of Apple Watch (Apple Watch Edition). Were you one of them? On April 10, 2015, when preorders first went on sale in nine countries, the $20,000 gold Apple Watch Edition model sold out in China in less than an hour.

>> **Apple Watch Hermès:** A partnership between Apple and Hermès, this fashion-centric watch includes bold, colorful (and extra-long wraparound) leather bands, and a new watch face you can't access anywhere else.

See `www.apple.com/watch` for more information on the Apple Watch collections and some accessories.

And once you've got an Apple Watch on your wrist, you've got several watch faces to choose from, all of which can be personalized to your liking. Here are just a few choices:

>> **Astronomy:** An out-of-the-world view of our solar system

>> **Chronograph:** Like an analog stopwatch

>> **Color:** Classic analog face with customizable colors

>> **Gradient:** Bright and bold colors that change throughout the day.

>> **Mickey or Minnie Mouse:** A classic returns — and in animation

>> **Modular Compact:** Bold digital watch face with lots of options

>> **Motion:** Animated objects, such as butterflies and flowers

>> **Numerals Mono:** A stylish hybrid of digital and analog time

>> **Simple:** A minimalistic but stylish analog watch

>> **Solar Dial:** Based on your location and time of day, you can see the sun's position

>> **Utility:** Analog watch with optional calendar reminders and more

>> **X-Large:** Large digital font for viewing at a distance

See Figure 13-4 for watch face examples, and see Chapter 4 for more on all the watch faces available to you and how to customize them.

DID YOU KNOW?

Although it was an unproven product for Apple, Apple Watch was a hit right out of the gate. According to Rakuten Intelligence (`intelligence.slice.com`), an online shopping analytics firm with a panel of more than two million online shoppers in the United States, 62 percent of Apple Watches were sold in the first hour of availability on April 10, 2015; 86 percent sold by 8 a.m. Approximately 1,039,000 people bought 1.4 million Apple Watches. As for which of the three collections sold the most, 62 percent bought Apple Watch Sport (least expensive offering), 38 percent purchased Apple Watch, and less than 1 percent ordered Apple Watch Edition (with an average price of $11,000 apiece).

FIGURE 13-4:
Choose from Astronomy, Mickey Mouse, or several other watch faces to personalize your Apple Watch.

Gaming

It's a huge understatement to say Apple Watch is an unproven video game platform. But given Apple's track record with iOS devices — not to mention a passionate app development community eager to take advantage of this new real estate on the wrist — gaming might be the secret "killer app" of Apple Watch.

You're in line at the supermarket and you want to kill some time by dunking a few virtual baskets by tapping your watch screen. Or you're on the train to work and you want to use your fingertip to slide letter tiles on a board to create a word. Or perhaps you're walking down the street and you feel a tap on your wrist — an alert that someone is invading your village and you've got to decide what to do.

FIGURE 13-5:
A look at Rules! for Apple Watch — based on the popular iOS version.

Just as the smartphone and tablet have become viable gaming platforms in a very short period of time — even pumping out such iconic games as Angry Birds and Flappy Bird — Apple Watch could introduce fresh gaming experiences on a device we always have strapped to our wrists.

Chapter 12 offers a look at a half-dozen games available for Apple Watch, but the App Store — accessible on the companion Apple Watch app on iPhone — has thousands more to choose from. Figure 13-5 shows what Rules! — a popular iOS game — looks like on the Apple Watch (and, yes, it's available for both platforms).

Index

A

About option for Apple Watch, 304
accelerometer, 29, 39
accessibility features, 54–55, 305–306
achievements, 233–234
Activity app
  benefits, 336–337
  Exercise ring, 218, 223–224, 336
  feature overview, 16–17
  icon, 21
  initial display, 217–218
  iPhone, 234–236
  Move ring, 218–222, 336
  overview, 74
  settings, 306
  with Siri, 206
  Stand ring, 218, 224–225, 336
Activity complication, 114
Activity (Wheelchair) option, 55
activity watch faces, 88–89
Airplane mode, 304
AirPlay, 259
Alarm app, 13, 21, 72, 118
Alarm complication, 113
Albums, 257
ambient light sensor, 30
American Airlines app, 311, 312
animojis, 150
App Layout, 303
App Store, 88, 297
App Store app, 21, 81
Apple ID, 305
Apple News app, 78
Apple Pay
  buying instructions, 337–338
  feature overview, 16
  overview, 77
  PIN requirement, 39–40
  setting up, 279–281
  Side button activation, 63, 64

Apple Pay app, 23
Apple Pay Cash, 288–292
Apple Store app, 21
Apple TV, 77, 273–275
Apple Watch. *See also* apps
  accelerometer, 29, 39
  accessibility features, 54–55, 305–306
  accessing time, 114–116
  ambient light sensor, 30
  Apple TV connectivity, 77, 273–275
  available countries, 12
  band, 26, 34, 83
  box, 34
  built-in apps, 21–23, 70–83
  bumpers, 50–51
  Calendar, 14
  caller ID, 13
  Cellular, 28
  cellular connectivity on, 41–43
  collections, 10–12
  continuity, defined, 28
  Control Center, 66–67
  controls, 58–64
  copying photos to, 318–319
  Dock, 14, 68–69, 169–171
  durability of, 50
  ease of use, 57–58
  ECG monitor, 245
  electrocardiogram (ECG), 30
  Emergency SOS, 13, 251–252
  Fall Detection, 249–251
  features overview, 12–18
  fitness features, 16–17, 23, 30, 75, 206, 215–217, 225–234, 243–252. *See also* Activity app
  GPS, 29
  hands-free, 64–65
  haptic technology, 31
  hardware elements, 25–30, 34, 39, 43–46, 49–54, 63–64, 83, 113, 114, 216, 243–247, 249–252. *See also* Digital Crown button; watch faces

Apple Watch *(continued)*

   Heart Rate app, 247–249

   Home screen, 19, 20–24, 26–27, 46–49

   icons, 21–23

   internal sensors, 27–30, 216, 243–245

   introduction, 1–6

   iTunes connectivity, 16, 273

   locations available, 12

   maintaining, 49–54

   Maps, 340–341

   media, 16, 77, 253–275. *See also* Music app

   Medical ID information, 246–247

   mirroring iPhone, 304

   Music, 340

   My Watch tab, 303–308

   Notifications, 69, 171–175, 303

   operating system requirement, 35

   orientation, 305

   pairing with iPhone, 35–39

   parts, 24–26

   responsible use of, 51–54

   setting up, 34–46

   Siri, 16, 63–65, 73, 118, 128, 140–141, 178, 180, 189, 196–200, 202–213, 259–260, 342–343

   sizes, 10

   storage capacity, 58, 264

   taps, 12, 26, 144, 146–147, 171–175

   time-related features, 13, 21, 23, 62, 72, 114–122, 202–204, 307–308

   touchscreen, 26–27

   Walkie-Talkie, 13, 338–339

   wireless function, 27–28, 124, 130–132, 260–262

Apple Watch app on iPhone, 19, 21–24, 29, 48, 298–302

Apple Watch Bumper, 50–51

Apple Watch Edition, 344, 345

Apple Watch For Dummies Cheat Sheet (website), 5

Apple Watch Hermès, 11

Apple Watch Nike+, 11

Apple Watch Series 3, 10

Apple Watch Series 4, 11

Apple Watch Series 5, 11

Apple Watch User Guide (website), 88

apps. *See also specific apps*

   built-in, 21–23, 70–83

   choosing, 48

downloading, 298–302

   games, 324–332, 346–347

   recommended, 308–315

AQI, UVI complication, 114

Archive (website), 272

Artists, Music app, 256–257

Astronomy watch face, 89

audio playback, battery life test, 44. *See also* Music app

audiobooks, 269–272

Audiobooks app, 21, 80

Automatic Downloads option, 305

B

back sensors, 26

band, 26, 34, 83

battery

   accessories for charging, 45–46

   back sensors/charger, relationship to, 26

   included power equipment, 34

   monitoring battery life, 43–45

   overview, 18

   Power Reserve mode, 45, 113

Battery Life (website), 46

Best Fiends game, 327

Bluetooth 4.0

   overview, 27–28

   pairing, 260–262

   receiving calls, 124, 130–132

Bluetooth low energy (BLE), 293

BMW i Remote app, 311

Bold Text option, 54, 306

Breathe app, 21, 78, 238–240

Breathe watch face, 89–90

Brightness & Text Size, 303

bumpers, defined, 50–51

buttons

   Digital Crown, 25, 48, 49, 62–65, 75, 89, 91, 109, 115–117, 120, 121, 336, 341

   Play, 256

   Power/Side, 25, 63–64, 114

   Reset, 122

   Side/Power, 25, 63–64, 114

   Start, 120, 122

   Stop, 120

C

calculating numbers with Siri, 211
Calculator app, 21, 82, 182–183
Calendar app, 14, 21, 72, 176–180, 206
Calendar complication, 113
California watch face, 90–91
caller ID, 13
calls
  incoming, 124, 125–126, 130–132
  introduction, 123–124
  outgoing, 126–128
Camera app, 22, 76, 318, 321–323
Cellular, 28
cellular connectivity, 41–43
charging/charger
  accessories for charging, 45–46
  back sensors/charger, relationship to, 26
  included power equipment, 34
  monitoring battery life, 43–45
  Power Reserve mode, 45, 113
Chimes option, 55
Chirp for Twitter app, 310–311
Chronograph watch face, 91
Citymapper app, 312–313
Clock app. *See also* World Clock app
  accessing, 114–117
  Digital Crown control of, 62
  icon, 21
  settings, 307–308
  with Siri, 202–204
C|Net (website), 5
CNN app, 312
collections, 10–12
Color watch face, 91, 92
communicating by using Apple Watch
  calls, 123–128, 130–132
  Digital Touch, 14, 59–62, 144–148, 341
  email (Mail app), 14, 22, 71, 154–159, 205–206
  text messages (Messages app), 14, 22, 70–71, 133–137, 139–142, 204, 308
Compass app, 21, 83, 189–191
complications. *See specific complications*
Contacts app, 78, 204–205
Contacts from iPhone, viewing, 78
Continuity, defined, 28
Control Center, 66–67

controls
  Digital Crown button, 25, 48, 49, 62–65, 75, 89, 91, 109, 115–117, 120, 121, 336, 341
  gestures, 26–27, 58–62. *See also* taps
  Side/Power button, 25, 63–64, 114
  Siri, 63, 64–65
  Sketch, 61–62
Cook, Tim, 107
copying photos to Apple Watch, 318–319
Cosmos Rings game, 332
credit cards, 77, 278
customizations compared to complications, 110–114
Cycle Tracking app, 22, 82, 240–242

D

debit cards, 77, 278
Device Account Number (DAN), 280, 284
Digital Crown button
  Astronomy face changes with, 89
  Clock access, 115
  controls, 62–63
  face color changing with, 91
  as Home screen return, 48
  multitasking with, 336
  overview, 25
  photo controls, 75
  Siri launch with, 49, 64–65
  Sketch, 341
  Stopwatch access, 120
  Timer access, 121
  twisting to customize, 109
  World Clock access, 116–117
digital handshake, defined, 28, 278
Digital Touch
  defined, 14
  Heartbeat, 60–61, 145, 148, 341
  Sketch, 61–62, 144, 145–146, 341
  Tap, 144, 146–147
  taps, relationship to, 59
Do Not Disturb tab, 304
Dock, 14, 68–69, 169–171
double-tap, 59
downloading apps, 298–302
durability feature, 50

E

eBay app, 312
electrocardiogram (ECG), 30, 245
email (Mail app)
  feature overview, 14
  icon, 22
  overview, 71
  receiving and managing, 154–159
  reply limitation of, 158
  with Siri, 205–206
Emergency SOS, 251–252
Emergency SOS feature, 13
emojis, 136–137
Enable Handoff option, 305
Engadget (website), 5
ESPN app, 309
Evernote app, 311
Exercise ring, 218, 223–224, 336
Explorer watch face, 92
Extra Large Watch Face option, 54, 306

F

faces, watch
  activity, 88–89
  Apple Watch, 13, 24
  Astronomy, 89
  Breathe, 89–90
  built-in, 88–106
  California, 90–91
  choosing, 107–110
  Chronograph, 91
  Color, 91, 92
  complications, 113, 114, 168–169
  customizations, 110–114, 168–169
  examples, 344–346
  Explorer, 92
  extra large option, 54, 306
  Fire and Water, 92–93
  Gradient, 94
  Infograph Modular, 88, 94–95
  Kaleidoscope, 95
  Liquid Metal, 96
  Mickey/Minnie Mouse, 96–97
  Modular, 97–98
  Motion, 98, 99

  Numerals Duo/Numerals Mono, 98–100
  Photos, 100
  Pride, 101
  Simple, 101, 102
  Siri, 101, 102, 201–202
  Solar Dial, 103
  Time Lapse, 103–104
  Toy Story, 104
  Utility, 104–105
  Vapor, 105, 106
  X-Large, 105, 106
FaceTime audio, 128–130
Fall Detection, 249–251
Fandango app, 314
Find Friends app, 22, 79
Find My Friends app, 212
Find My iPhone feature, 280
Fire and Water watch face, 92–93
fitness features
  achievement tracking, 233–234
  Activity app, 16–17, 21, 74, 206, 217–225,
    234–236, 308, 336–337
  Breathe app, 238–240
  Cycle Tracking app, 240–242
  ECG monitor, 245
  Emergency SOS, 251–252
  Fall Detection, 249–251
  Heart Rate app, 247–249
  heart rate sensor, 30, 216, 243–245
  introduction, 215–217
  Medical ID information, 246–247
  Noise app, 242
  overview, 16–17
  reminders, 232
  summary, 233
  Trends app, 236–237
  Workout app, 16–17, 23, 75, 206, 225–231
Font Adjustment option, 54, 306
Force Touch, 27, 59–60
friends, 153, 212

G

games
  Best Fiends, 327
  Cosmos Rings, 332

possibilities for, 346–347

Rules!, 329, 331

Runeblade, 325–326

Snappy Word, 327–328

Spy_Watch, 328–329, 330

Trivia Crack, 325, 326

Watch Quest, 328, 329

Watch This Homerun, 324–325

Wordie, 330–331

General options, My Watch, 304–305

gestures

  press, 27, 59–60

  swipe, 27, 61

  taps, 12, 26, 144, 146–147, 171–175

  two-finger press, 60–61

Gizmodo (website), 5

global positioning system (GPS), 29

GPS, 18–19

GPS + Cellular, 19

Gradient watch face, 94

Grayscale option, 54, 306

H

handoff feature, 130–131, 305

hands-free, 64–65

haptic technology, 31, 174, 303, 306

hardware elements

  accelerometer, 29, 39

  band, 26, 83

  battery, 26, 43–46, 113

  Digital Crown button, 25, 48, 49, 62–65, 75, 89, 91, 109, 115–117, 120, 121, 336, 341

  ECG monitor, 30, 245

  Emergency SOS, 251–252

  Fall Detection, 249–251

  maintaining, 49–54

  Medical ID information, 246–247

  sensors, 27–30, 216, 243–245

  Side/Power button, 25, 63–64, 114

  touchscreen, 26–27

  watch face. *See* watch faces

  watchband, 26, 34, 83

Health option, 304

Heart Rate app, 22, 79, 247–249

heart rate sensor, 30, 216, 243–245

Heartbeat, 60–61, 145, 148, 341

heartbeats, sending, 342

Hey, Siri option, 196

History tab, 235

Home app, 22, 82

Home screen

  introduction, 20–24

  iPhone, 19

  tap, 26

  working with, 46–49

hotel key feature, 338

I

iBeacon (website), 292

iCloud (website), 39, 280

icons. *See specific icons*

iMessage, 70, 133

incoming calls, 124, 125–126, 130–132

induction technology, 46

Infograph Modular watch face, 88, 94–95

information access

  Calculator app, 182–183

  Calendar app, 176–180

  Compass app, 189–191

  Dock, 169–171

  introduction, 161–162

  Maps app, 184–189

  Notifications, 69, 171–175, 303

  Reminders app, 181–182

  Stocks app, 74, 165–169

  Voice Memos app, 183–184

  Weather app, 23, 73, 162–165

internal sensors, 27–30, 216, 243–245

Internet connection for Siri, 65

iPhone

  Activity app, 234–236

  alarm synchronization, 118

  Apple Pay, 279–281, 282–283

  Apple Pay app, 277

  Apple Watch app, 19, 21–24, 29, 48, 298–302

  Breathe app, 238–240

  calls, 123–128, 130–132

  camera link to iPhone, 76

  choosing apps, 46–49

  connecting Apple Watch to, 65

iPhone *(continued)*

Contacts, viewing with Apple Watch, 78

Cycle Tracking app, 240–242

Find My iPhone feature, 280

GPS, 29

health and fitness support, 215–217

Home screen icon for Apple Watch, 19

listening to music through, 260–262

mirroring, 304

Noise app, 242

operating system requirements, 35

pairing with, 35–39

PIN, 39

Siri options, 196–200

Trends app, 236–237

iTunes, 16, 273

iTunes Radio, 77

K

Kaleidoscope watch face, 95

Kanex GoPower Watch Plus, 46

L

Language & Region options, 305

languages for Siri, 197

Last Used App option, 115

LibriVox (website), 271

linear actuator, defined, 65

Liquid Metal watch face, 96

location, sending, 143

location information, 29

Long-Look Notification, 69

Loop (website), 5

Loyal Books (website), 271

Lutron Caséta app, 314

M

Mail app

feature overview, 14

icon, 22

overview, 71

receiving and managing, 154–159

reply limitation of, 158

with Siri, 205–206

maintenance, 49–54

Maps app

icon, 22

navigating with, 184–189, 340–341

overview, 16, 75, 76, 340–341

with Siri, 189, 206–207

media

Apple TV, 273–275

audiobooks, 269–272

introduction, 253–254

iTunes, 16, 273

Music app, 76–77, 207–209, 254–266, 308, 340

podcasts, 266–269

radio plays, 272–273

video remote control, 77, 273–275

Medical ID information, 246–247

memojis, 150

Messages app

deleting conversations, 143–144

emojis in, 136–137

icon, 22

overview, 14, 70–71

receiving, 133–136

responding, 133–136

Scribbles feature in, 136–137

sending, 139–142

settings, 308

with Siri, 140–141, 204

Mickey/Minnie Mouse watch face, 96–97

Mint app, 308

Mirror iPhone option, 304

mobile payments with Apple Pay

buying instructions, 337–338

feature overview, 16

overview, 77

PIN requirement, 39–40

setting up, 279–281

without nearby iPhone, 284–285

Modular watch face, 97–98

Mono Audio option, 55, 306

Moon Phase complication, 113

Motion watch face, 98, 99

Motor option, 55

Move ring, 218–222, 336

movement, measuring your, 29

multitasking, 336

Music, 340

Music app
  AirPlay, 259
  Albums, 257
  Artists, 256–257
  listening to music through iPhone, 260–262
  managing music from iPhone, 254–259
  Now Playing, 256
  overview, 76–77
  playback, 262–265, 340
  Playlists, 258–259
  removing music, 265–266
  settings, 308
  with Siri, 207–209, 259–260
  Songs, 258

My Information tab, 198–199

My Watch tab, 303–308

N

navigating with Apple Watch (Compass app),
  189–191

navigating with Apple Watch (Maps app)
  icon, 22
  navigating with, 184–189, 340–341
  overview, 16, 75, 76
  with Siri, 189, 206–207

near field communication (NFC), 28, 63

netiquette, 53

News app, 22

9to5Mac (website), 5

Noise app, 22, 81, 242

Nomad Pod Pro, 45

Notifications
  accessing, 171–175
  activating, 69
  changing settings, 303

long-look, 69
  short-look, 69

Notifications Center, 174

Now Playing app, 22, 256

NPR One app, 313

numbers, calculating with Siri, 211

Numerals Duo/Numerals Mono watch face, 98–100

O

OneDrive app, 309

On/Off Label options, 54, 306

OpenTable app, 311

operating system requirement, 35

options, 305–307

orientation, watch, 305

OTRCat (website), 272

outgoing calls, 126–128

P

pairing via wireless connection, 35–39, 260–262

passcode (PIN) for Apple Watch, choosing, 39–40

Passcode option, 304

PayByPhone Parking app, 314

Phone app, 22, 70, 204–205

phone calls
  incoming, 124, 125–126, 130–132
  introduction, 123–124
  outgoing, 126–128

photoplethysmography, 243–244

photos, taking with Camera, 21, 76, 318, 321–323

Photos app, 12, 75, 318–321

Photos Limit option, 319

Photos watch face, 100

PIN for Apple Watch, 39–40

ping, defined, 67

place shifting, defined, 267

Play button, 256

playing
  audiobooks, 269–272
  music, 262–265, 340
  podcasts, 266–269

Playlists, 258–259

plays, radio, 272–273

podcasts, 266–269

Podcasts app, 23, 79

power equipment (battery)

   accessories for charging, 45–46

   back sensors/charger, relationship to, 26

   included power equipment, 34

   monitoring battery life, 43–45

   Power Reserve mode, 45, 113

Power Reserve mode, 45, 113

Power/Side button, 25, 63–64, 114

preannouncement, defined, 306

pressing, 27, 59–60

Pride watch face, 101

privacy and Apple Pay, 280

Project Gutenberg (website), 271

Prominent Haptic option, 174, 306

R

Radio app, 23, 80

radio plays, 272–273

raising wrist to activate feature, 116

reading texts with Siri, 211

receiving

   emails, 154–159

   text messages, 133–136

Reduce Motion option, 54, 306

Reduce Transparency option, 54, 306

region setting, 305

reminders, 210–211, 232

Reminders app, 23, 181–182

Remote app, 23, 77, 273–275

removing music, 265–266

Reset button, 122

responding to messages, 133–136

rings

   Exercise, 218, 223–224, 336

   Move, 218–222, 336

   Stand, 218, 224–225, 336

Roeding, Cyriac, 292–293

Rules! game, 329, 331

Runeblade game, 325–326

S

S1, defined, 27

safety

   Apple Pay, 280

   driving cautions, 125

   passcode (PIN), 39–40, 304

Saltzman, Marc (author)

   Siri For Dummies, 3, 16, 196, 335

screenshot, taking, 169

Scribble option, 55

Scribbles feature, 137–138

scrolling with Digital Crown button, 63

Secure Element, defined, 280

security

   adding credit/debit card to Apple Pay, 280

   options for, 307

   passcode (PIN), 39–40, 304

   theft of watch, preventing, 53–54

sending

   heartbeats, 148, 342

   location, 143

   sketches, 145–146

   taps, 144, 146–147

   text messages, 139–142

sensors, 27–30, 216, 243–245

Settings app, 23, 83, 174, 261

shopBeacon app, 293

shopkick app, 292–293

Short-Look Notification, 69

Side/Power button, 25, 63–64, 114

Simple watch face, 101

Siri

   Activity app, 206

   alarm setting with, 118

   benefits of using, 342–343

   calculating numbers, 211

   Calendar app, 178, 180, 206

   Clock/World Clock apps, 202–204

   Contacts app, 204–205

   controls for, 64–65

   Digital Crown button activation, 63

   finding friends, 212

   hands-free, 64–65

   having fun with, 212–213

Hey, Siri option, 196

Internet connection for, 65

languages, 197

Mail app, 205–206

making calls with, 128, 204–205

Maps app, 189, 206–207

messaging with, 140–141, 204

Music app, 207–209, 259–260

My Information tab, 198–199

overview, 16, 73

reading texts, 211

reminders, 210–211

setting up, 196–200

talking to, 199–200

voice feedback, 198

Workout app, 206

Siri For Dummies (Saltzman), 3, 16, 196, 335

Siri watch face, 101, 102, 201–202

Sketch, 144, 145–146

Sky Guide app, 314, 315

smartwatch (Apple Watch). *See* Apple Watch; apps

SMS text message, 70. *See also* Messages app

Snappy Word game, 327–328

Software Update option, 304

Solar Dial watch face, 103

Songs, 258

Sounds & Haptics, 303

SPG (Starwood Preferred Guest), 310, 338

Spy_Watch game, 328–329, 330

Stand ring, 218, 224–225, 336

Start button, 120, 122

Start Pairing tab, 35

Starwood Preferred Guest (SPG) app, 310, 338

stickers, 149–150

Stocks app, 74, 165–169

Stocks complication, 113, 168–169

Stop button, 120

Stopwatch app, 23, 72, 118–121

Stopwatch complication, 113

storage capacity, 58, 264

Sunrise/Sunset complication, 113

swipe, 27, 61

Synced Album option, 318

T

talk time, battery life test, 44

Tap category, Digital Touch, 144, 146–147, 341

tapback, 150–151

tapping, touchscreen, 26, 58–59

taps

defined, 12, 26, 144

Notifications, 171–175

sending, 144, 146–147

Taptic Engine, 31, 55, 65

Target app, 309

temperature extremes, avoiding, 50

Text, Bold, 54, 306

text messages. *See* Messages app

theft of watch, preventing, 53–54

Things app, 314

time accuracy (website), 107

Time Lapse watch face, 103–104

time shifting, defined, 267

timer, camera, 76

Timer app, 13, 72, 121–122

Timer complication, 113

time-related features

accessing time, 114–117

Alarm app, 2, 13, 21, 118

Clock app, 21, 62, 114–117, 202–204, 307–308

Stopwatch app, 23, 72, 118–121

Timer app, 13, 72, 121–122

World Clock app, 72, 116–117, 202–204

Touch Accommodations option, 55

Touch ID sensor, 280

touchscreen, 26–27

Toy Story watch face, 104

Trends app, 236–237

TripAdvisor app, 313

Trivia Crack game, 325, 326

two-finger press, 60–61

U

USB port, 43–44

Utility watch face, 104–105

V

Vapor watch face, 105, 106
viewfinder, Apple Watch as, 76
voice dictation, 135
Voice Feedback from Siri, 198
Voice Memos app, 23, 82, 183–184
Voicemail from iPhone, accessing, 126
VoiceOver option, 54, 59, 306

W

Walkie-Talkie app, 13, 23, 78, 152–154, 338–339
Wallet app, 23, 77, 285–288
watch, battery life test, 45
watch faces. *See* faces, watch
Watch Orientation option, 305
Watch Quest game, 328, 329
Watch This Homerun game, 324–325
watchband, 26, 34, 83
water, avoiding, 24, 49, 226
Weather app, 23, 73, 162–165
Weather complication, 113, 168–169
websites. *See specific websites*

Wi-Fi, 28, 124
Wikipedia (website), 5
wireless function
    Bluetooth 4.0, 27–28, 124, 130–132, 260–262
    Wi-Fi, 28, 124
Wordie game, 330–331
workout, battery life test, 45
Workout app, 5, 16–17, 23, 206, 225–231
World Clock app, 72, 116–117, 202–204
World Clock complication, 113
World Time, accessing, 116–117
Wrist Detection option, 305
Wrist Raise, 116

X

X-Large watch face, 105, 106

Z

Zoom option, 54, 306
zooming with Digital Crown button, 63

About the Author

Marc Saltzman is a prolific journalist, author, and TV/radio personality who specializes in consumer electronics, business technology, interactive entertainment, and Internet trends. Marc has authored 16 books since 1996 and currently contributes to more than 25 high-profile publications, including *USA TODAY/Gannett*, Yahoo!, MSN, *AARP: The Magazine*, Common Sense Media, *Costco Connection*, *Toronto Star*, Zoomer, and others. Marc hosts various video segments, including "Gear Guide" (seen at Cineplex movie theaters across Canada), and is the host of the Tech Impact television show, seen on Bloomberg TV. Marc also hosts "Tech It Out," a nationally syndicated radio show (and podcast) via Radio America. *Apple Watch For Dummies* is Marc's second Dummies book. He also penned *Siri For Dummies*.

Follow Marc on Twitter: @marc_saltzman. Or visit his website at www. marcsaltzman.com

Dedication

This book is dedicated to my extraordinary parents, Stan and Honey Saltzman. A heartfelt thank you for your never-ending support, guidance, and love.

Author's Acknowledgments

I'd like to acknowledge all the talented folks at John Wiley & Sons, for this book wouldn't have happened without their professionalism, patience, and knowledge (hardly Dummies!). In particular, I'd like to thank Matt Fecher, as well as awesome acquisitions editor Ashley Barth.

Publisher's Acknowledgments

Executive Editor: Ashley Barth

Development Editor: Rebecca Senninger

Technical Editor: Matthew Fecher

Production Editor: Magesh Elangovan

Cover Image: Courtesy of Marc Saltzman

Leverage the power

Dummies is the global leader in the reference category and one of the most trusted and highly regarded brands in the world. No longer just focused on books, customers now have access to the dummies content they need in the format they want. Together we'll craft a solution that engages your customers, stands out from the competition, and helps you meet your goals.

Advertising & Sponsorships

Connect with an engaged audience on a powerful multimedia site, and position your message alongside expert how-to content. Dummies.com is a one-stop shop for free, online information and know-how curated by a team of experts.

- Targeted ads
- Video
- Email Marketing
- Microsites
- Sweepstakes sponsorship

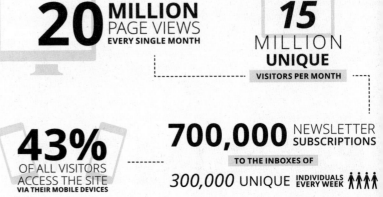

20 MILLION PAGE VIEWS EVERY SINGLE MONTH

15 MILLION UNIQUE VISITORS PER MONTH

43% OF ALL VISITORS ACCESS THE SITE VIA THEIR MOBILE DEVICES

700,000 NEWSLETTER SUBSCRIPTIONS TO THE INBOXES OF

300,000 UNIQUE INDIVIDUALS EVERY WEEK

# of dummies

Custom Publishing

Reach a global audience in any language by creating a solution that will differentiate you from competitors, amplify your message, and encourage customers to make a buying decision.

- Apps
- Books
- eBooks
- Video
- Audio
- Webinars

Brand Licensing & Content

Leverage the strength of the world's most popular reference brand to reach new audiences and channels of distribution.

For more information, visit **dummies.com/biz**

PERSONAL ENRICHMENT

Staying Sharp dummies
9781119187790
USA $26.00
CAN $31.99
UK £19.99

Facebook dummies
Carolyn Abram
9781119179030
USA $21.99
CAN $25.99
UK £16.99

Guitar dummies
Mark Phillips
Jon Chappell
9781119293354
USA $24.99
CAN $29.99
UK £17.99

Investing dummies
Eric Tyson, MBA
9781119293347
USA $22.99
CAN $27.99
UK £16.99

Beekeeping dummies
Howland Blackiston
9781119310068
USA $22.99
CAN $27.99
UK £16.99

Digital Photography dummies
Julie Adair King
9781119235606
USA $24.99
CAN $29.99
UK £17.99

Meditation dummies
Stephan Bodian
9781119251163
USA $24.99
CAN $29.99
UK £17.99

Pregnancy ALL-IN-ONE dummies
9781119235491
USA $26.99
CAN $31.99
UK £19.99

Samsung Galaxy S7 dummies
Bill Hughes
9781119279952
USA $24.99
CAN $29.99
UK £17.99

iPhone dummies
Edward C. Baig
Bob "Dr. Mac" LeVitus
9781119283133
USA $24.99
CAN $29.99
UK £17.99

Crocheting dummies
Karen Manthey
Susan Brittain
9781119287117
USA $24.99
CAN $29.99
UK £16.99

Nutrition dummies
Carol Ann Rinzler
9781119130246
USA $22.99
CAN $27.99
UK £16.99

PROFESSIONAL DEVELOPMENT

Windows 10 dummies
Andy Rathbone
9781119311041
USA $24.99
CAN $29.99
UK £17.99

AutoCAD dummies
Bill Fane
9781119255796
USA $39.99
CAN $47.99
UK £27.99

Excel 2016 dummies
Greg Harvey, PhD
9781119293439
USA $26.99
CAN $31.99
UK £19.99

QuickBooks 2017 dummies
Stephen L. Nelson, MBA, CPA, BS in Taxation
9781119281467
USA $26.99
CAN $31.99
UK £19.99

macOS Sierra dummies
Bob "Dr. Mac" LeVitus
9781119280651
USA $29.99
CAN $35.99
UK £21.99

LinkedIn dummies
Joel Elad, MBAs
9781119251132
USA $24.99
CAN $29.99
UK £17.99

Windows 10 ALL-IN-ONE dummies
Woody Leonhard
9781119310563
USA $34.00
CAN $41.99
UK £24.99

SharePoint 2016 dummies
Rosemarie Withee
Ken Withee
9781119181705
USA $29.99
CAN $35.99
UK £21.99

Fundamental Analysis dummies
Matt Krantz
9781119263593
USA $26.99
CAN $31.99
UK £19.99

Networking dummies
Doug Lowe
9781119257769
USA $29.99
CAN $35.99
UK £21.99

Office 2016 dummies
Wallace Wang
9781119293477
USA $26.99
CAN $31.99
UK £19.99

Office 365 dummies
Rosemarie Withee
Ken Withee
Jennifer Reed
9781119265313
USA $24.99
CAN $29.99
UK £17.99

Salesforce.com dummies
Liz Kao
Jon Paz
9781119239314
USA $29.99
CAN $35.99
UK £21.99

Coding dummies
Nikhil Abraham
9781119293323
USA $29.99
CAN $35.99
UK £21.99

dummies.com

dummies
A Wiley Brand

Learning Made Easy

ACADEMIC

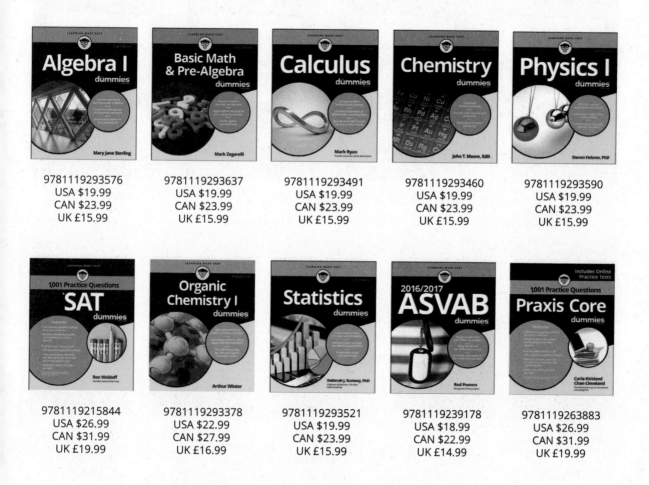

Algebra I dummies
Mary Jane Sterling

9781119293576
USA $19.99
CAN $23.99
UK £15.99

Basic Math & Pre-Algebra dummies
Mark Zegarelli

9781119293637
USA $19.99
CAN $23.99
UK £15.99

Calculus dummies
Mark Ryan

9781119293491
USA $19.99
CAN $23.99
UK £15.99

Chemistry dummies
John T. Moore, EdD

9781119293460
USA $19.99
CAN $23.99
UK £15.99

Physics I dummies
Steven Holzner, PhD

9781119293590
USA $19.99
CAN $23.99
UK £15.99

1,001 Practice Questions SAT dummies
Ron Woldoff

9781119215844
USA $26.99
CAN $31.99
UK £19.99

Organic Chemistry I dummies
Arthur Winter

9781119293378
USA $22.99
CAN $27.99
UK £16.99

Statistics dummies
Deborah J. Rumsey, PhD

9781119293521
USA $19.99
CAN $23.99
UK £15.99

2016/2017 ASVAB dummies
Rod Powers

9781119239178
USA $18.99
CAN $22.99
UK £14.99

1,001 Practice Questions Praxis Core dummies
Carla Kirkland
Chan Cleveland

9781119263883
USA $26.99
CAN $31.99
UK £19.99

Available Everywhere Books Are Sold

dummies.com

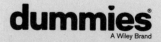

dummies®
A Wiley Brand

Small books for big imaginations

GETTING STARTED WITH Coding
Get Creative with Code!
Camille McCue, PhD

9781119177173
USA $9.99
CAN $9.99
UK £8.99

MODDING Minecraft
Build Your Own Minecraft Mods!
Sarah Guthals, PhD
Stephen Foster, PhD
Lindsey Handley, PhD

9781119177272
USA $9.99
CAN $9.99
UK £8.99

MAKING YouTube VIDEOS
Star in Your Own Video!
Nick Willoughby

9781119177241
USA $9.99
CAN $9.99
UK £8.99

DESIGNING Digital Games
Create Games with Scratch!
Derek Breen

9781119177210
USA $9.99
CAN $9.99
UK £8.99

GETTING STARTED WITH Raspberry Pi
Program Your Raspberry Pi!
Richard Wentk

9781119262657
USA $9.99
CAN $9.99
UK £6.99

EXPERIMENTING WITH Science
Think, Test, and Learn!
Olivia J. Mullins, PhD

9781119291336
USA $9.99
CAN $9.99
UK £6.99

CREATING Digital Animations
Animate Stories with Scratch!
Derek Breen

9781119233527
USA $9.99
CAN $9.99
UK £6.99

GETTING STARTED WITH Engineering
Think Like an Engineer!
Camille McCue, PhD

9781119291220
USA $9.99
CAN $9.99
UK £6.99

WRITING Computer Code
Learn the Language of Computers!
Chris Minnick and Eva Holland

9781119177302
USA $9.99
CAN $9.99
UK £8.99

Unleash Their Creativity

dummies.com

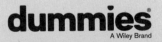